AF818035

Sustainable Global Development in Environment, Energy, Building, and Infrastructure

Md. Faruque Hossain

Sustainable Global Development in Environment, Energy, Building, and Infrastructure

Springer

Md. Faruque Hossain
Argonne National Laboratory
Lemont, IL, USA

ISBN 978-3-031-84428-7 ISBN 978-3-031-84429-4 (eBook)
https://doi.org/10.1007/978-3-031-84429-4

© The Editor(s) (if applicable) and The Author(s), under exclusive license to Springer Nature Switzerland AG 2025

This work is subject to copyright. All rights are solely and exclusively licensed by the Publisher, whether the whole or part of the material is concerned, specifically the rights of translation, reprinting, reuse of illustrations, recitation, broadcasting, reproduction on microfilms or in any other physical way, and transmission or information storage and retrieval, electronic adaptation, computer software, or by similar or dissimilar methodology now known or hereafter developed.
The use of general descriptive names, registered names, trademarks, service marks, etc. in this publication does not imply, even in the absence of a specific statement, that such names are exempt from the relevant protective laws and regulations and therefore free for general use.
The publisher, the authors and the editors are safe to assume that the advice and information in this book are believed to be true and accurate at the date of publication. Neither the publisher nor the authors or the editors give a warranty, expressed or implied, with respect to the material contained herein or for any errors or omissions that may have been made. The publisher remains neutral with regard to jurisdictional claims in published maps and institutional affiliations.

This Springer imprint is published by the registered company Springer Nature Switzerland AG
The registered company address is: Gewerbestrasse 11, 6330 Cham, Switzerland

If disposing of this product, please recycle the paper.

To
The future captain who will build a better planet

Preface

For the past hundreds of years, conventional applications of science and engineering for the development of the global environment, power, shelter, infrastructure, and transportation sectors (EPSIT) have put our mother Earth in a vulnerable condition. Just because these conventional development mechanisms in our daily lives are destroying our natural resources and environmental equilibrium rapidly. They are presently required by nature to be "environmentally friendly" to protect the earth for our future generations. Therefore, it is time to act now individually and/or commonly to apply science and engineering principles in every sector of EPSIT in order to secure a better world.

In this book, thus, a holistic approaches of sustainability application in all sectors of the EPSIT system has been discussed by presenting a series of innovative research, and its possible applications for building a sustainable world for our future generation. Naturally, this book defines the context of instigation of intelligent thinking of scientific theories and its practical applications to confirm global sustainability through the implementation of the following six themes, Part I: Introduction; Part II: Sustainable Environment; Part III: Sustainable Energy; Part IV: Sustainable Building; Part V: Sustainable Infrastructure and Part VI: Conclusion which are very much interconnected to secure a better planet.

Lemont, IL, USA Md. Faruque Hossain

Competing Interests The author has no competing interests to declare that are relevant to the content of this manuscript.

Contents

Part I Introduction

1 Global Environmental Vulnerability . . . 3
Introduction . . . 3
Materials and Methods . . . 4
CO_2 Emission . . . 4
Global Health Crisis . . . 4
Climate Change . . . 5
Results and Discussion . . . 5
CO_2 Emission . . . 5
Global Health Crisis . . . 6
Climate Change . . . 6
Conclusion . . . 6
References . . . 8

Part II Sustainable Environment

2 The Origin of Global Climate Change . . . 13
Introduction . . . 13
Materials and Methods . . . 14
Results and Discussion . . . 16
Conclusion . . . 18
References . . . 18

3 Climate Change Impacts on Global Ecosystem . . . 21
Introduction . . . 21
Materials and Methods . . . 22
Mechanistic Climate Change . . . 22
Risk on Global Ecosystem . . . 23

Results and Discussion 23
Mechanistic Climate Change 23
Risk on Global Ecosystem 24
Conclusions 26
References 27

4 Remodeling of Photosynthesis to Equilibrate Global CO_2 Cycle . . . 29
Introduction 29
Methods and Simulation 29
Modeling of Electric Field Energy (E_e) 29
Modeling of Photon Frequency (f) and Wavelength (λ) 32
Results and Discussion 32
Modeling of Electric Field Energy (E_e) 32
Modeling of Photon Frequency (f) and Wavelength (λ) 34
Conclusion 37
References 37

5 Application of Dormant Light Energy to Breakdown CO_2 Naturally 39
Introduction 39
Methods and Simulation 39
Modeling of Dormant Solar Energy (D_e) 39
Activation of Dormant Light Energy (D_e) to Break the CO_2 42
Results and Discussion 44
Modeling of Dormant Light Energy (D_e) 44
Activation of Dormant Light Energy (D_e) to Break the CO_2 47
Conclusions 49
References 50

Part III Sustainable Energy

6 The Availability of Global Solar Energy 55
Introduction 55
Materials and Methods 55
Estimation of Total Solar Energy Reaching on Earth 55
Calculation of Total Electricity Energy from Total Solar Radiation on the Earth 57
Results and Discussion 59
Estimation of Total Solar Energy Reaching the Earth 59
Calculation of Total Electricity Energy from Total Solar Radiation on the Earth 62
Conclusion 65
References 65

7 Remodeling of Photon Structure to Produce Renewable Energy . . . 69
Introduction 69

Methods and Simulations 69
Results and Discussion 71
Conclusions 75
References 76

8 Natural Mechanism of Electricity Generation from the Atmospheric Water Molecules 79
Introduction 79
Materials and Methods 80
Electrostatic Force Generation 80
Photoelectrolysis for Energy Formation 81
Results and Discussion 83
Electrostatic Force 83
Photoelectrolysis for Energy Formation 84
Conversion of H_2 Energy into Electricity 85
Conclusions 86
References 87

9 Modeling of the Intelligence of a Tree to Generate Ultimate Energy 91
Introduction 91
Methods and Simulation 92
Simulation of Cognitive Power (C_p) of a Tree 92
Simulation of Physical Energy (P_e) of a Tree 94
Computation of Environmental Energy (Ee) 96
Results and Discussion 100
Simulation of Cognitive Power (C_p) of a Tree 100
Simulation of Physical Energy (P_e) of a Tree 102
Computation of Environmental Energy (E_e) 103
Conclusions 107
References 107

Part IV Sustainable Building

10 Application of Solar Energy in Design Technology to Naturally Cool and Heat the Houses 113
Introduction 113
Methods and Simulation 114
Cooling Mechanism 114
Heating Mechanism 115
Results and Discussion 117
Cooling Mechanism 117
Heating Mechanism 119
Conclusions 122
References 124

11 Application of Printable Solar Cell in the Building Design to Mitigate Net Energy Demand Naturally 127
Introduction 127
Methods and Materials 128
Results and Discussion 132
Conversion of Electricity 133
Savings on Energy Cost 135
Conclusions 135
References 136

12 Implementation of Sonic Reaction Technology in Building Design to Eliminate Pathogens Naturally 139
Introduction 139
Materials and Methods 140
HuGF Formation by the Interior Building Wall 140
Killing Pathogens 141
Results and Discussion 142
HuGF Formation by the Interior Building Wall 142
Killing Pathogens 144
Conclusions 145
References 146

Part V Sustainable Infrastructure

13 Intransitive Technology of the Plant's Transpiration Mechanism to Meet the Global Water Perplexity 151
Introduction 151
Materials and Methods 152
Static Electric Force Formation 152
In Site Water Treatment 154
Results and Discussion 155
Electrostatic Force Analysis 155
Conclusions 156
References 156

14 Simulation of the Environmental Force for Development of Global Infrastructure System 159
Introduction 159
Methods and Simulation 160
Earth's Electromagnetic Force 160
Invisible Infrastructure 163
Air Dynamics Transport 164
Results and Discussion 167
Earth's Electromagnetic Force 167
Invisible Infrastructure 169
Air Dynamics Transport 171

Conclusions 172
References 172

Part VI Conclusion

15 The Modeling of Global Environmental Equilibrium 177
Introduction 177
Methods and Simulation 178
Simulation of the Phototrophic Energy (Pt_e) 178
Implementation of Phototrophic Energy (Pt_e) to Break the CO_2 180
Simulation of the Chemotrophic Mechanism (Ct_m) 182
Chemotrophic Mechanism to Eliminate Pollutants 184
Results and Discussion 185
Simulation of the Phototrophic Energy (Pt_e) 185
Implementation of Phototrophic Energy (Pt_e) to Break the CO_2 187
Simulation of the Chemotrophic Mechanism (Ct_m) 190
Chemotrophic Mechanism to Eliminate Pollutants 191
Conclusions 193
References 193

Index 197

About the Author

Md. Faruque Hossain has more than 20 years of industry experience in the field of sustainability research, development, and project management under global top public agencies and fortune-listed companies. He worked and/or consulted in diverse global top tier companies to conduct research and development for million dollars to over billion dollars projects for ensuring global sustainability. Faruque also worked for the NYC Department of Citywide Administrative Services as Senior Management team and interacted with the heads of all public agencies, highest level government officials of local, state, federal, and international organization leaders for developing global sustainability policy. During his tenure in NYC Department of Environmental Protection as Acting Director, Faruque managed a world-class team of scientists, consultants, architects, engineers, and contractors from AECOM, Fluor, Skanska, and maintained highest level professional relationship to conduct global sustainability research and practice. Hossain received his Ph.D. from Hokkaido University, did post-graduate research in Chemical Engineering at the University of Sydney, and Executive Education in Architecture at Harvard University. He is an LEED-certified professional and editors of several International Journal of Global Sustainability-related field. Dr. Hossain is renowned as the industry leader and notable scientist to conduct innovative research and project development for energy, environment, building, infrastructure sustainability field for building a better Earth. He has hundreds of world-class publications in very high impact journals, and he wrote six books (published by Springer, Elsevier) and eighty-two book chapters (published by Springer, Elsevier, and Francis & Taylor) in the field of global sustainability. Currently, he is working at the Argonne National Laboratory, IL, USA as a long-term visiting scholar with his motto to practice sustainability globally for building a better planet.

Part I
Introduction

Chapter 1
Global Environmental Vulnerability

Introduction

Owing to economic, political, social, and demographic crises, people are trying to move to a better place to have a better quality of life [4, 5]. However, political barriers make this goal more difficult for people who want to relocate to a more advantageous location [1, 9, 18]. Subsequently, people are needed to embed to settle in the specific area that creates densely populated areas. Subsequently, they build conventional homes and infrastructure for their shelters, which leads to release of huge amounts of carbon dioxide emissions, thus, global health crisis and climate change occur massively that make them more susceptible to have a better life on this planet [2, 13]. Hence, this study tried to analyze the adverse effects on human ecology by applying integrated assessment model (IAM) including explicit CO_2 emissions, global health issues, and the climate change challenge in relation to political barriers. Subsequently, the quantification impact of political hurdles on population density and its role in CO_2 emissions, global health issues, and climate change on a worldwide scale was also analyzed. Simply in this study, the current impact on population dynamics has been analyzed in relation to the political barrier between nations considering the quantitative data of CO_2 emission, global health crisis, and climate change scenarios by implementing IAM empirical model [6, 10]. Thus, the overall goal of this work was to use IAM to analyze the assessment and quantification of CO_2 emissions, global health, and climatic change caused by conventional development and population density on a worldwide scale to give an early warning to overcome these problems to achieve a cleaner, greener earth for our survival.

© The Author(s), under exclusive license to Springer Nature Switzerland AG 2025

M. F. Hossain, *Sustainable Global Development in Environment, Energy, Building, and Infrastructure*, https://doi.org/10.1007/978-3-031-84429-4_1

Materials and Methods

In this study, an integrated assessment model (IAM) was employed from the data Department of Environmental Protection (DEP), United States Department of Energy (USDOE), Intergovernmental Panel on Climate Change (IPCC), Carbon Foundation Council (CFC), Carbon Dioxide Information and Analysis Center (CDIAC), International Energy Agency (IEA), United Nations Environmental Program (UNEP), National Environmental Association Agency (NEAA), New Energy and Industrial Technology Development Organization (NEDO), National Oceanic and Atmospheric Administration (NOAA), and National Aeronautics and Space Administration (NASA) to illuminate the effects of international political obstacles on global health and climate change. In addition, gravity modeling has been used to analyze whether global political barrier patterns are becoming more similar or more different, considering both the population dynamics and their traditional history.

CO_2 Emission

The escalation of CO_2 emissions due to interstate political impediments has been elucidated by a population dynamics model that incorporates conventional development factors. Thus, the evaluation of environmental complexity has been elucidated by the demonstration of its direct correlation with population dynamics. This was accomplished through the implementation of the inaugural CO_2 emission robustness model derived from conventional development. The traditional development parameter in this model assesses long-run equilibria influencing vulnerability to CO_2 emissions in relation to population dynamics [11, 17]. Therefore, this study estimates the global CO_2 emissions using IAMs to assess the magnitude of air pollution coefficients linked to national population trends. Hence, the political barriers between nations generate significant population dynamics, the primary factor affecting these coefficients; thus, the objective of simulating total CO_2 emissions stemming from population dynamics is to elucidate the impact of political obstacles. This was accomplished by employing IAM to incorporate a squared term related to origin or population dynamics to facilitate the examination of how political impediments influence global health crisis and climate change.

Global Health Crisis

By incorporating the impact of CO_2 emissions using IAM into the gravity estimation with ordinary least squares (OLS), the parameters of the global health problem have been analyzed to get the impact on global health due to the CO_2 emissions. Consequently, the total global health problem is calculated by calibrating the CO_2

emissions rate and considering the population dynamics caused by international boundaries [3, 16]. These characteristics have been transferred from the worldwide health problem to the concurrently occurring CO_2 emissions on a global scale [8, 12]. Subsequently, a demographic model of population dynamics in connection to environmental change was also used to generate projections of population dynamics to clarify the scenarios involving population dynamics. Then the gravity model was implemented to clarify the CO_2 emissions on population dynamics to identify the impact of it on the climate change.

Climate Change

Incorporating IAM into the gravity estimate using ordinary least squares (OLS) is being utilized to examine the issue of global climate change, which is fueled by CO_2 emissions generated by political barriers between states. These barriers substantially contribute to emitting billions of tons of carbon dioxide annually [7, 15]. Hence, the modeling of carbon cycle and the climate's responsiveness to CO_2 emissions are consequently analyzed to formulate the global carbon budget to compute the global carbon budget annually with respect to the influence of political obstacles among the nations to monitor rapid changes in CO_2 levels in the atmosphere.

Results and Discussion

The calculation evidence from the IAM of time series of nonstationary CO_2 emissions worldwide in relation to the global health and climate change suggested that the planet is in a grave danger condition. Hence, the research suggested that the abiotic (from the hydrosphere, lithosphere, and atmosphere) and biotic (from living things and microbes) pollution have both significantly changed the dynamics of their surrounding environments due to the political barriers among the nations and are largely responsible for the global health and climate change crisis.

CO_2 Emission

The research suggests that densely populated areas are the main factor for massive CO_2 emission as per IAM data interpreted from the data DEP, USDOE, IPCC, CFC, CDIAC, IEA, UNEP, NEAA, NEDO, NOAA, and NASA. Naturally, the atmospheric concentration of carbon dioxide has been steadily rising since 1960s, which means that these emissions are either floating there or have the potential to travel great distances, contaminating the atmospheric environment in the process [7, 14]. Hence, calculating the rate of rise in atmospheric CO_2 concentration over a ten-year period involves comparing decadal and individual annual values suggesting that from 1880 to 2029, the yearly increase in global CO_2 levels is 0.19 gigatonnes in the 1960s, 0.28 gigatonnes in the 1970s, 0.14 gigatonnes in the 1980s, 0.72 gigatonnes in the 1990s, 1.48 gigatonnes in the 2000s, 1.48 gigatonnes in the 2010s, and 1.48 gigatonnes in the 2020s. The results point to the fact that it is clear and immediate danger to our planet as it increases annual global CO_2 emissions by 1.418% in recent years (*see* Table 1.1).

Global Health Crisis

The study results suggest that the release of carbon dioxide into the atmosphere is the root cause of the worldwide health crisis. Hence, the IAM data interpreted from DEP, USDOE, IPCC, CFC, CDIAC, IEA, UNEP, NEAA, NEDO, NOAA, and NASA suggesting that this emission significantly damages the human respiratory system and causes bronchitis, mucus production, zoonoses, cholera, bronchospasm, cardiovascular disease, mental, typhoid, malaria, perinatal disorders, and neuron of cell death (*see* Fig. 1.1). This finding indicates that the global danger to human health is being substantially exacerbated by global CO_2 emissions since it has direct consequences on the global human health around the world.

Climate Change

The IAM suggested that the planet's natural process to sequester carbon dioxide is unsaturated level, and thus, its influence on climate changes. This result demonstrates that the influence of population dynamics on climate change is the primary factor contributing to the degradation of nature's capacity to absorb CO_2, serving as an indicator of climate change exposure and susceptibility, subsequently leading to global warming, which considerably impacts the tipping points that have risen by 0.89 °C over the past few decades [3, 8]. Consequently, it is the primary cause of melting of polar ice caps, acid rains, rising sea levels, flooding, and all global natural disasters that disrupt the planet's environment equilibrium.

Conclusion

This study presents the results of a quantitative investigation of the effects of political barriers on global public health and climate change vulnerability. Hence, the findings of this study were derived from the qualitative data of DEP, USDOE, IPCC, CFC, CDIAC, IEA, UNEP, NEAA, NEDO, NOAA, and NASA, incorporated into the IAM modeling which suggested that the most significant factor that affects

Table 1.1 Results from the integrated assessment model (IAM) simulation show the net rise of CO_2 (GtC Year^{-1}) emissions from 1880 to 1999, 1990 to 1919, 1920 to 1939, 1940 to 1959, 1960 to 1969, 1970 to 1979, 1980 to 1989, 1990 to 1999, 2000 to 2009, 2010 to 2019, and 2020 to 2029

Mean (GtC Year^{-1})	1880–1999	1900–1919	1920–1939	1940–1959	1960–1969	1970–1979	1980–1989	1990–1999	2000–2009	2010–2019	2020–2029
Total CO_2 emission globally											
Total CO_2 emission globally: IAM simulations corresponding to the mean average data interpretation of DEP, USDOE, IPCC, CFC, CDIAC, IEA, UNEP, NEAA, NEDO, NOAA, and NASA	0.40 ± 0.20	0.50 ± 0.20	0.60 ± 0.10	1.60 ± 0.10	1.70 ± 0.20	1.70 ± 0.20	1.60 ± 0.10	2.60 ± 0.10	2.60 ± 0.10	3.26 ± 0.20	3.26 ± 0.20
	0.40	0.50	0.60	1.60	1.70	1.70	1.60	2.60	2.60	3.26	3.26
	0.60	0.70	0.70	1.70	1.90	1.90	1.70	2.70	2.70	3.46	3.46
	0.20	0.30	0.50	1.60	1.70	1.50	1.50	2.60	2.50	3.06	3.06
Average global CO_2 emission	**0.40**	**0.50**	**0.60**	**1.63**	**1.77**	**1.70**	**1.60**	**2.63**	**2.60**	**3.26**	**3.26**
Total CO_2 sequestration globally											
Total CO_2 sequestration globally: IAM simulations corresponding to the mean average data interpretation of DEP, USDOE, IPCC, CFC, CDIAC, IEA, UNEP, NEAA, NEDO, NOAA, and NASA	1.30 ± 0.10	1.30 ± 0.10	1.40 ± 0.20	1.60 ± 0.20	1.50 ± 0.10	1.30 ± 0.10	1.40 ± 0.20	1.60 ± 0.20	1.15 ± 0.20	1.15 ± 0.10	1.15 ± 0.10
	1.50	1.30	1.40	1.60	1.50	1.30	1.40	1.60	1.15	1.15	1.15
	1.60	1.40	1.60	1.80	1.60	1.40	1.60	1.80	1.35	1.25	1.25
	1.40	1.20	1.20	1.40	1.40	1.20	1.20	1.40	1.95	1.05	1.05
Average CO_2 sequestration globally	**1.50**	**1.30**	**1.40**	**1.60**	**1.50**	**1.30**	**1.40**	**1.60**	**1.48**	**1.15**	**1.15**
Annual Total growth rate of atmospheric CO_2											
Total annual increasing rate of CO_2	**−1.10**	**−0.80**	**−0.80**	**0.03**	**0.27**	**0.40**	**0.20**	**1.03**	**1.12**	**2.11**	**2.11**
Annual increasing rate of CO_2	**−0.77**	**−0.56**	**−0.56**	**0.02**	**0.19**	**0.28**	**0.14**	**0.72**	**0.78**	**1.48**	**1.48**

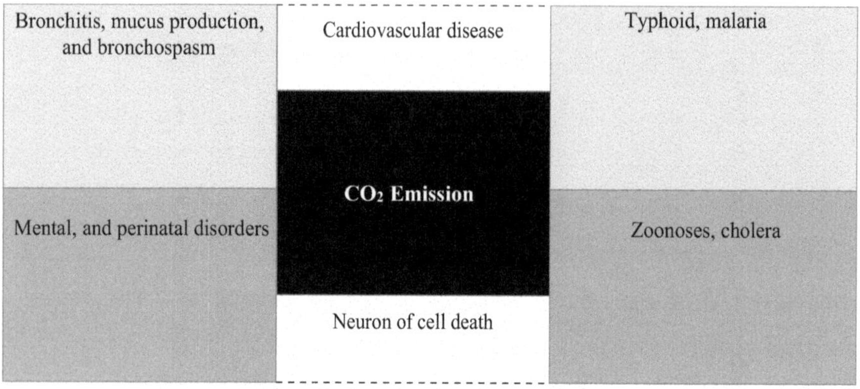

Fig. 1.1 Shows the worldwide infectious diseases as an ongoing concern for public health due to the impact of massive CO_2 emission generated by the political barrier

the global health and climate change crisis are the restrictive political barrier among the countries. Subsequently, the global public health risk and environmental equilibrium conundrum are in great danger that need to be addressed immediately to sustain any living being on the earth in the near future.

Acknowledgments The author is fully credited throughout the document for all the findings, forecasts, and discoveries that are described in this chapter.

References

1. Achard, F., Beuchle, R., Mayaux, P., Stibig, H., Bodart, C., Brink, A., & Eva, H. D. (2014). Determination of tropical deforestation rates and related carbon losses from 1990 to 2010. *Global Change Biology, 20*(8), 2540–2554.
2. Bauer, J. E., Cai, W., Raymond, P. A., Bianchi, T. S., Hopkinson, C. S., & Regnier, P. A. (2013). The changing carbon cycle of the coastal ocean. *Nature, 504*(7478), 61–70.
3. Benveniste, H. (2021). *Science for international policy on climate change: A multi-method contribution*. Princeton University.
4. Chevallier, F. (2015). On the statistical optimality of CO_2 atmospheric inversions assimilating CO_2 column retrievals. *Atmospheric Chemistry and Physics, 15*(19), 11133–11145.
5. Davis, S. J., & Caldeira, K. (2010). Consumption-based accounting of CO_2 emissions. *Proceedings of the National Academy of Sciences, 107*(12), 5687–5692.
6. Hélène, B. O., & Marc, F. (2020). Effect of border policy on exposure and vulnerability to climate change. *Proceedings of the National Academy of Sciences*.
7. Hossain, M. F. (2018). Global environmental vulnerability and the survival period of all living beings on earth. *International Journal of Environmental Science and Technology*.
8. Hossain, M. F. (2021). *Global sustainability in energy, building, infrastructure, transportation, and water technology*. Springer Science and Business Media LLC.
9. Hossain, M. F. (2022). Extreme level of CO_2 accumulation into the atmosphere due to the unequilibrium global carbon emission and sequestration. *Water, Air and Soil Pollution, 233*, 105.

10. Hossain, M. F. (2022). Sustainable design for global equilibrium., Springer Science and Business.
11. Hossain, M. F. (2023). *Global sustainability*. Springer Science and Business Media LLC.
12. Houghton, R. A. (2007). Balancing the global carbon budget. *Annual Review of Earth and Planetary Sciences, 35*(1), 313–347.
13. Le Quéré, C., Andres, R. J., Boden, T., Conway, T., Houghton, R. A., House, J. I., Marland, G., Peters, G. P., van der Werf, G. R., Ahlström, A., Andrew, R. M., Bopp, L., Canadell, J. G., Ciais, P., Doney, S. C., Enright, C., Friedlingstein, P., Huntingford, C., Jain, A. K., Jourdain, C., Kato, E., Keeling, R. F., Klein Goldewijk, K., Levis, S., Levy, P., Lomas, M., Poulter, B., Raupach, M. R., Schwinger, J., Sitch, S., Stocker, B. D., Viovy, N., Zaehle, S., & Zeng, N. (2013). The global carbon budget 1959–2011. *Earth System Science Data*.
14. Li, W., Ciais, P., Wang, Y., Peng, S., Broquet, G., Ballantyne, A. P., & Le Quéré, C. (2016). Reducing uncertainties in decadal variability of the global carbon budget with multiple datasets. *Proceedings of the National Academy of Sciences, 113*(46), 13104–13108.
15. Rather, S.-U., Sulaimon, A. A., Shariff, A. M., Qasim, A., et al. (2023). Investigation of physical properties and carbon dioxide solubility in tetramethylammonium bromide and tetraethylammonium bromide ionic liquids solutions. *Chemosphere*.
16. Schwietzke, S., Sherwood, O. A., Bruhwiler, L. M., Miller, J. B., Etiope, G., Dlugokencky, E. J., & White, J. W. (2016). Upward revision of global fossil fuel methane emissions based on isotope database. *Nature, 538*(7623), 88–91.
17. Stephens, B. B., Gurney, K. R., Tans, P. P., Sweeney, C., Peters, W., Bruhwiler, L., & Nakazawa, T. (2007). Weak northern and strong tropical land carbon uptake from vertical profiles of atmospheric CO_2. *Science, 316*(5832), 1732–1735.
18. Wang, Z., & Liu, T. (2015). Scenario analysis of regional carbon reduction targets in China: A case study of Beijing. *Journal of Renewable and Sustainable Energy*.

Part II
Sustainable Environment

Chapter 2
The Origin of Global Climate Change

Introduction

The enormous development of industrial sector (manufacturing, mining, construction, and agriculture) and transportation sector (vehicles that transport people or goods, such as cars, trucks, buses, motorcycles, trains, aircraft, boats, barges, and ships) around the world is quickening to the accumulation of air CO_2 that is indeed an alarming sign for human beings to take quality air to breathe soon [3, 7]. Ultimately, revealing nature's contributions to adopt this deadly element produced by the conventional development of the building infrastructure account for massive CO_2 emissions would be the worst scenarios soon and transforming it to a world in which both nature and people will face severe environmental and health crisis soon [2, 5]. Naturally, it is imperative to prioritize the establishment of sustainable global environment practices at this juncture to ensure the resilience, adaptability, and effective management of our planet, without causing any adverse effects that our future generations can have better environment to take fresh air [1, 12].

Therefore, in this work, a spatially explicit modeling of the total CO_2 emission to air due to the contribution has been conducted on two vital conventional development factors of building sector and infrastructure sector. Consequently, these dual-dimensional contributions to nature have been analyzed considering nature's current and nature's future CO_2 concentration and distinguishing these contributions to the atmosphere and human well-being. Simply, an advancement of complete calculation of global CO_2 emission from the two primary contributors of building and the infrastructure sector has been conducted. Subsequently, the sequestration of CO_2 by ocean and land vegetation has been calculated by DVGM simulation to determine the accurate increasing rate of CO_2 into the air to prepare humankind for impending environmental danger in the near future. Naturally, this work will certainly assist the scientists, engineers, architects, developers, and leaders throughout the world to realize this imminent danger earnestly to console global CO_2 emissions instantaneously by enforcing sustainable development technology in the building and

© The Author(s), under exclusive license to Springer Nature Switzerland AG 2025

M. F. Hossain, *Sustainable Global Development in Environment, Energy, Building, and Infrastructure*, https://doi.org/10.1007/978-3-031-84429-4_2

infrastructure sectors to address the imminent severe air quality threat for the mankind to survive on the earth soon [4, 8]. So, it is time to think about the societal awareness of the importance of environmentally friendly sustainable development in building and infrastructure sectors for the benefit of mankind well-being throughout the world.

Materials and Methods

The calculation of CO_2 emissions in the air resulting from the conventional development of the industrial sector (IS) is conducted using dynamic global environmental modeling (DGVM) simulations. This analysis considers the variations between two consecutive decades, specifically the 1960s, 1970s, 1980s, 1990s, 2000s, 2010s, and 2020s. The conversion process involved dividing the annual emissions of the current year by the emissions of the previous year, as expressed by the following equation:

$$IS = \left[\frac{E_{IS(t_0+9)} - E_{IS(t_0)}}{E_{IS(t_0)}}\right] \times 100\% \text{year}^{-1} \tag{2.1}$$

Here, E represents the CO_2 emission, IS represents building sector, t_0 represents the first year of the decade, t_9 represents the following 9 years of the decade, and year^{-1} represents the per year CO_2 emission. This present analysis pertains to the examination of a basic computation to ascertain the annual rate of increase in carbon dioxide emissions. To accurately establish the rate at which CO_2 is increasing over each decade, a leap-year factor (IS_{LP}) is incorporated into the calculation [9, 13]. This factor is utilized to compute the correct rate of CO_2 emissions by employing the following logarithmic function:

$$IS_{LP} = \frac{1}{E_{IS}} \frac{d\,(\ln E_{IS})}{dt} \tag{2.2}$$

Similarly, the annual growth rate of carbon dioxide emissions resulting from the expansion of the conventional development of transportation sector (TS) on a global scale was determined by comparing the emissions of two consecutive decades spanning from the 1960s to the 2020s. This growth rate was subsequently converted into a yearly percentage by dividing the difference in emissions between the previous year and the current year by the emissions of the previous year, as expressed in the following equation:

$$TS = \left[\frac{E_{TS(t_0+9)} - E_{TS(t_0)}}{E_{TS\,(t_0)}}\right] \times 100\% \text{year}^{-1} \tag{2.3}$$

As the equation presented accounts for the annual growth rate of CO_2 emissions resulting from conventional infrastructure development, and thus, accurately reflects the net increase in carbon dioxide (TS_{LP}), a leap-year component is incorporated by stating the equation as follows:

$$TS_{LP} = \frac{1}{E_{TS}} \frac{d\left(\ln E_{TS}\right)}{dt} \tag{2.4}$$

Then, the global net CO_2 emission considering 10-year period for each experimental data has been conducted, and then, it has been transformed into a representation of the average annual rate considering the period of 1960s to 2020s by using the similar modeling as described earlier [10, 13]. Simply, the rates of CO_2 emissions have been computed for various time scales, ranging from decades, by utilizing nonlinear equations to determine the annual CO_2 emissions. Consequently, the total CO_2 emissions resulting from the conventional development of all building and infrastructure sectors each year have been determined by combining the results obtained from four equations (Eqs. 2.1, 2.2, 2.3 and 2.4) as follows:

$$\begin{aligned} IS + TS = & \left(\left[\frac{E_{IS(t_0+9)} - E_{IS(t_0)}}{E_{BD\ (t_0)}}\right] \times 100\% \text{year}^{-1}\right) + \left(\frac{1}{E_{IS}} \frac{d\left(\ln E_{IS}\right)}{dt}\right) \\ & + \left(\left[\frac{E_{TS(t_0+9)} - E_{TS(t_0)}}{E_{TS\ (t_0)}}\right] \times 100\% \text{year}^{-1}\right) + \left(\frac{1}{E_{TS}} \frac{d\left(\ln E_{IS}\right)}{dt}\right) \end{aligned} \tag{2.5}$$

Subsequently, the comprehensive assessment of net carbon dioxide sequestration has been undertaken, encompassing both the ocean and the land vegetation sinks on a global scale [6, 15]. This analysis spans the period from the 1960s to the 2020s, with each dataset representing a 10-year interval. Then the resulting data has been transformed into an average yearly rate. This analysis has been conducted based on a series of decades, specifically the 1960s, 1970s, 1980s, 1990s, 2000s, 2010s, and 2020s by implementing a total carbon sink cycle in all oceans [10, 14]. The present methodology is being employed to precisely examine the mechanisms of worldwide oceans that are directly engaged in the sequestration of CO_2 by its flora and fauna [6, 11]. Hence, the precise determination of the carbon dioxide absorption capacity of all oceans is achieved by dividing the annual values by the corresponding values of the preceding year. Consequently, the annual rate of CO_2 sequestration by the oceans (t) in giga tons of carbon per year (GtC year^{-1}) is computed using the following formula:

$$S_{\text{OCEAN}}\ (t) = \frac{1}{n} \sum\nolimits_{m=1}^{m=n} \frac{S^{m}_{\text{OCEAN}}\ (t)}{S^{m}_{\text{OCEAN}(t10-t1)}} \tag{2.6}$$

In this context, the variable n denotes the quantity of oceans, m indicates the parameters associated with carbon dioxide sequestration, and t signifies the duration under consideration.

Then the annual absorption of carbon dioxide by terrestrial vegetation is currently being assessed to determine the overall carbon dioxide sequestration by land (S_{LAND}). This assessment involves examining data from the 1960s, 1970s, 1980s, 1990s, 2000s, 2010s, and 2020s, and converting it into an annual rate. In this analysis, the net carbon dioxide absorption by terrestrial ecosystems is being computed using the following methodology:

$$S_{\mathrm{LAND}} = E_{IS} + E_{TS} - (G_{\mathrm{ATM}} + S_{\mathrm{OCEAN}}) \tag{2.7}$$

Here, S_{LAND} is computed based on the remaining estimates, where G_{ATM} represents the presence of CO_2 into the air at a concentration of 400 ppm, E_{IS} represents the carbon that results from the development of the building sector, and E_{ID} represents the CO_2 that results from the development of the infrastructure sector [2, 11]. Hence, the determination of E_{ID} is accomplished by subtracting (G_{ATM} + S_{OCEAN}) CO_2 from the result of the computation of S_{LAND} in Equation [7].

Naturally, the total CO_2 sequestration over the course of a year was determined by combining these two equations (Eqs. 2.6 and 2.7), which resulted in the following calculation:

$$S_{\mathrm{OCEAN}}(t) + S_{\mathrm{LAND}} = \frac{1}{n}\sum_{m=1}^{m=n} \frac{S^{m}_{\mathrm{OCEAN}}(t)}{S^{m}_{\mathrm{OCEAN}^{(t10-t1)}}} + E_{IS} + E_{TS} - (G_{\mathrm{ATM}} + S_{\mathrm{OCEAN}}) \tag{2.8}$$

Finally, the net annual growth rate of CO_2 concentration in the atmosphere is calculated each year based on variations in total CO_2 emission and total CO_2 sequestration per year.

Results and Discussion

The data from 1960 to 2029 indicates that the average yearly worldwide CO_2 emissions in the air during this period amounted to an annual average of 10.77 GtC year^{-1} (giga ton carbon per year) in the years of 1960s (1960–1969), whereas 10.70 GtC year^{-1} per 10 years in the 1970s (1970–1979). Subsequently, the total global CO_2 emissions at a yearly average of 10.60 GtC year^{-1} in the years of 1980s (1980–1989); yearly average of 20.63 GtC year^{-1} in the years of 1990s (1990–1999), whereas 20.60 GtC year^{-1} in the 2000s (2000–2009) and yearly average of 34.26 GtC year^{-1} for the years of 2010s (2010–2019) and it projected; annual average of 34.26 GtC year^{-1} for the years of 2020s (2020–2029).

Subsequently, the total CO_2 that has been sank by the ocean and plants suggested that the average global CO_2 sink from 1960 to 2029 during this time scale throughout the world at an annual average of 10.50 GtC year^{-1} for the year of 1960s (1960–1969); annual average of 10.30 GtC year^{-1} for the years of 1970s

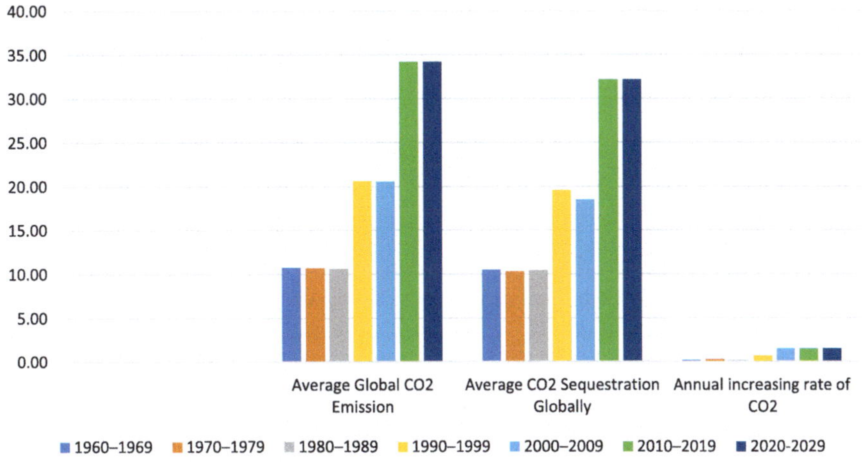

Fig. 2.1 The average annual growth in global CO_2 from 1960 to 2029 where it was 0.19 GtC year^{-1} in the 1960s, 0.28 GtC year^{-1} in the 1970s, 0.14 GtC year^{-1} in the 1980s, 0.72 GtC year^{-1} in the 1990s, 1.48 GtC year^{-1} in the 2000s, 1.48 GtC year^{-1} in the 2010s and is predicted to be 1.48 GtC year^{-1} in the 2020s and the projected growth rate CO_2 concentration will remain at 1.48 GtC year^{-1} (per year) for the next several decades

(1970–1979); annual average of 10.40 GtC year^{-1} for the year of 1980s (1980–1989); 19.60 GtC year^{-1} for the years of 1990s (1990–1999); and 18.48 GtC year^{-1} for the years of 2000s (2000–2009); 32.15 GtC year^{-1} for the years of 2010s (2010–2019), and it is projected 32.15 GtC year^{-1} for the years of 2020s (2020–2029) where 0.77 GtC year^{-1} (*see* Fig. 2.1).

Given the present atmospheric CO_2 concentration of 400 ppm, its annual increasing rate is 1.418 GtC year^{-1}, therefore, the CO_2 in the air will reach an unbreathable level of 1200 ppm in 75 years, as confirmed by the following calculation:

$$1200 = 400(1 + 0.0148)^{\text{Year}} \tag{2.9}$$

$$3 = (1 + 0.0148)^{\text{Year}} \tag{2.10}$$

$$\text{Log}\ (3) = \text{Year} \times \text{Log}(1.0148) \tag{2.11}$$

$$\text{Year} = \frac{\text{Log}\ 3}{\text{Log}\ 1.0148} \tag{2.12}$$

$$\text{Year} = 74.77 = 75\ (\text{roundup}) \tag{2.13}$$

Naturally, it is expected that the whole human population on the earth will experience significant respiratory difficulties resulting from this unbreathable level of CO_2 in the air that is originated initially from the conventional development building and infrastructure sectors globally. Hence, it is an immediate need to mitigate the emission of carbon dioxide by enforcing sustainable development technology globally in both building and infrastructure sectors instead of presently used conventional technology in these two sectors throughout the world to equilibrate the forthcoming severe respiratory problem for all humans in 75 years.

Conclusion

The global net carbon dioxide emissions resulting from the conventional development of the industry and transportation sectors worldwide have been estimated over the past few decades. Subsequently, the total CO_2 sequestration by oceans and land vegetation has also been calculated to verify the yearly rate of CO_2 increase that is 1.418 GtC $year^{-1}$ per year. As the present CO_2 level is 400 ppm in the air and it is increasing at a rate of 1.418 GtC $year^{-1}$, the CO_2 concentration in the air shall reach an unbreathable level of 1200 ppm in 75 years if the current practice of the conventional development of building and infrastructure system is not replaced and, thus, entire humans will face severe respiratory problem in the next 75 years. Simply, this research suggested that conventional development of industrial and transportation sectors is the primary cause for the climate change that must need to replace by sustainable development of these two sectors to curb CO_2 emission to secure fresh air for our next generation.

Acknowledgments The author asserts that all works of conceptualization, methodology, investigation, visualization, and writing presented in this chapter are the exclusive responsibility of the author.

References

1. Bauer, J. E., et al. (2013). The changing carbon cycle of the coastal ocean. *Nature, 504*(7478), 61–70.
2. Betts, R. A., et al. (2016). El niño and a record CO_2 rise. *Nature Climate Change, 6*(9), 806–810.
3. Davis, S. J., et al. (2010). Consumption-based accounting of CO_2 emissions. *Proceedings of the National Academy of Sciences, 107*(12), 5687–5692.
4. Erb, K., et al. (2013). Bias in the attribution of forest carbon sinks. *Nature Climate Change, 3* (10), 854–856.
5. Galina, C. (2013). *An introduction to carbon cycle science*. Cambridge University Press.
6. Hossain, M. F. (2022). Extreme level of CO_2 accumulation into the atmosphere due to the un-equilibrium global carbon emission and sequestration. *Water Air Soil Pollution, 233*, 105. (2022).
7. Houghton, R. A. (2007). Balancing the global carbon budget. *Annual Review of Earth and Planetary Sciences, 35*(1), 313–347.
8. Le Quéré, C. (2013). The global carbon budget 1959–2011. *Earth System Science Data, 32*, 44–49.

9. Li, W., et al. (2016). Reducing uncertainties in decadal variability of the global carbon budget with multiple datasets. *Proceedings of the National Academy of Sciences, 113*(46), 13104–13108.
10. Liu, Z., et al. (2015). Reduced carbon emission estimates from fossil fuel combustion and cement production in China. *Nature, 524*(7565), 335–338.
11. Mason, E. J., et al. (2012). Timing of carbon emissions from global forest clearance. *Nature Climate Change, 2*(9), 682–685.
12. Morimoto, S., et al. (2021). Spatio-temporal variations of the atmospheric greenhouse gases and their sources and sinks in the arctic region. *Polar Science, 27*, 100553.
13. Sami-Ullah, R., et al. (2023). Investigation of physical properties and carbon dioxide solubility in tetramethylammonium bromide and tetraethylammonium bromide ionic liquids solutions. *Chemosphere, 21*, 434–439.
14. Stephens, B. B., et al. (2007). Weak northern and strong tropical land carbon uptake from vertical profiles of atmospheric CO_2. *Science, 316*(5832), 1732–1735.
15. Wu, T., et al. (2013). Global carbon budgets simulated by the Beijing Climate Center climate system model for the last century: Global carbon cycle from a Chip5 model. *Journal of Geophysical Research Atmospheres, 25*, 234–239.

Chapter 3
Climate Change Impacts on Global Ecosystem

Introduction

The global ecosystem absorbs carbon, supports enormous terrestrial biodiversity, and secures the global environmental balance to protect the mother earth [2, 5]. Though their excellent carbon sequestration capacity to leverage the ecosystem, the future of global ecosystem is in danger due to its imperiled nature to sink the CO_2 to equilibrate the global environment [4, 7]. Simply, the global ecosystem faces substantial danger that can play vital role for carbon cycle feedback which indeed can accelerate to undermine their role in global CO_2 mitigation to make an environmental equilibrium nature on the earth [8, 10]. Subsequently, the risks to global vegetation stability and losses of biodiversity causing extreme weather including heightened death rates, species turnover, and physiological constraints to growth or regeneration are all potential challenges that could be encountered on global vegetation [1, 13].

Although several recent studies have been conducted to estimate the ecosystem carbon balance, the empirical assessment of ecosystem impact due to climate change has not been discussed yet [12, 19]. Hence, the Intergovernmental Panel on Climate Change (IPCC) confirms the risk of climate change as the eminent for the global ecosystem on their susceptibility [3, 14]. Thus, the extent of these studies of climate change risks on biodiversity has not been confirmed globally nor has been approached adequately to quantify these risks systematically. Subsequently, the extensive range of global biome structure of climate change danger to global ecosystem is not defined precisely. Thus, in this research, modeling for risk framework for global ecosystem, mechanistic ecosystem models, risk estimates from "climate envelope" species or ecoregion approaches, risk calculation from "climate change," and cross-approach analysis have been commuted to precisely understand the climate change impact on the global ecosystem.

© The Author(s), under exclusive license to Springer Nature Switzerland AG 2025

M. F. Hossain, *Sustainable Global Development in Environment, Energy, Building, and Infrastructure*, https://doi.org/10.1007/978-3-031-84429-4_3

Materials and Methods

Mechanistic Climate Change

Global climate change mechanics has been computed in taking everything into account of the distribution of Maxent bioclimatic variables on the global ecosystem for the period of 1923–2023. Then climate models from CMIP5 data archives have been incorporated to estimate the average net species presence in grid cells for the period of 1923–2023. Thus, the average ecosystem loss between historical 1923s and 2023s can be calculated considering a high emission of CO_2 scenario compared to mitigation of CO_2 scenario (typically RCP2.6). Then the aggregate climate change responses to mean changes per biome were estimated considering risk of ecosystem loss of each biome in the multimode mean of SSP3. All these estimates are being regridded to the common 1 × 1 degree grid considering the global climate change; thus, these estimates are being captured in a variable axis of risk than the CMIP6 models, confirming changes in ecosystem and species turnover and in the whole scale shifted biomes globally [2, 9]. Therefore, these climate change metrics are primarily calculated based on the risk associated with exposure, with a limited ability to account for adaptive capacity. Hence, the adaptive capacity is implicitly included only in the variation of current ecoregion distributions within each ecoregion globally [11, 15].

Subsequently, the analysis of climate sensitivity has been conducted considering stand-clearing disturbances to leverage the disturbance rate ecosystem globally with Landsat-derived stand-clearing disturbance rates globally [6, 16]. Naturally, the quadratic ordinary least calculation models, depending on the climate change sensitivities of disturbance, have been conducted to clarify the impact of biome disturbance regime [17, 18]. Then the use of the z-score anomalies in SSP3 from the ESMs has been estimated in stand-clearing disturbance on global ecosystem for the period 1923–2023. Consequently, a tested λ value has been analyzed to clarify the global stand-clearing ecosystem disturbances for the period 1923–2023 due to the climate change. Subsequently, a function of mean annual biome using the predictor variables of the multimode average of the ESMs has been analyzed from ERA5 historical data to compute net changes on global ecosystem due to the climate change climate.

This analysis is being considering the variables between two consecutive decades of the 1923s, 1933s, 1943s, 1953s, 1963s, 1973s, 1983s, 1993s, 2003s, 2013s, and 2023s. Then the conversion process has been conducted by dividing the annual climate change (M_{cl}) of the current year from the climate change of the previous year, as expressed by the following equation:

$$M_{cl} = \left[\frac{M_{cl\ (t_{0+9})} - M_{cl\ (t_0)}}{M_{cl\ (t_0)}} \right] \times 100\% \text{year}^{-1} \quad (3.1)$$

Here, t_0 represents first year of the decade and t_9 represents following 9 years of the decade and year^{-1} represents the per year climate change variable.

Risk on Global Ecosystem

Consequently, the broad risk assessment framework conducted by the integrated assessment models (IAMs) incorporation with the Intergovernmental Panel on Climate Change (IPCC) framework for global ecosystem (here carbon storage for climate mitigation) has been performed. The application of these risk structures to the global ecosystem presents a number of intrinsic issues, where the most significant one is the availability of data, in employing them to the global ecosystem [5, 7]. Therefore, the mechanistic ecosystem models that are contained in the earth system are being confirmed by the climate change models to identify the impact on global ecosystem.

Consequently, the dataset of the Coupled Model Intercomparison Project—Phase 6 (CMIP6) for all earth system models (ESMs) has been conducted to clarify the simulated changes in global ecosystem considering historical simulations for the period 1923–2023, shared socioeconomic pathway (SSP3) simulations for the period 1923–2023. Then, it was all regridded via bilinear implementation to a common 1×1 degree to then perform all models precisely for global ecosystem and calculate multimodel metric impacts. Hence, the global ecosystem dataset has been used to identify the direct impacts on global vegetation losses throughout the world. As this data portal is in the CMIP6 SSP3 simulation, thus, it can certainly confirm the precise interpretation of direct impact on global ecosystem due to the climate-driven changes. Therefore, each of the climate change models is being summed to clarify all primary and secondary impact on global ecosystem on each grid cell (Fig. 3.1). Subsequently, a further consistent spatial global ecosystem has been analyzed across all CMIP6 and empirical risk data portals to include only 1×1 degree grid cells in global ecosystem to ensure the impact of climate change on it.

Subsequently, the annual impact on global ecosystem (G_{EI}) resulting from climate change was determined by comparing the impacts of two consecutive decades spanning from the 1923s to the 2023s. This impact rate was subsequently converted into a yearly percentage by dividing the difference in impact between the previous years, as expressed in the following equation:

$$G_{EI} = \left[\frac{G_{EI\ (t_{0+9})} - G_{EI\ (t_0)}}{G_{EI\ (t_0)}}\right] \times 100\%\text{year}^{-1} \qquad (3.2)$$

Results and Discussion

Mechanistic Climate Change

The estimated risk of climate change in relation to the Maxent model projections (C risk axis) suggested that the climate sensitivities of global ecosystem are increasing in a pattern for the past decades due to the excessive amount of CO_2 in the air [8, 19]. Subsequently, the carbon concentration is the primary factor for climate change in consideration of high emission of CO_2 compared to mitigation of CO_2 in RCP2.6 [4, 5]. Subsequently, the risk metrics of climate change due to the excessive amount of CO_2 across the globe suggested that the climate change correlation matrix between the metrics of risk is higher equated with more risk in the environment. Here, the estimated risks across all datasets will represent the global climate change risk, and thus, these the CMIP6 model plotted distribution of percentiles for each metric suggested that climate change has a sever risk on global ecosystem [3, 5]. Subsequently, the metrics of the spatial of SSP3 confirm the changes in the composition of climate change and the metric of risk of climate change increasing globally which indeed a clear danger for the global ecosystem. Subsequently, the results from these impacts have been also compared to confirm a global impact of climate risks on ecosystem. Interestingly, the results suggested that the climate change in global environment from mechanistic models in respect to the Coupled Model Intercomparison Project Phase 6 (CMIP6) has the direct influences to change to contextualize overall global ecosystem. Thus, the global ecosystem risk through empirical climate envelope model results suggested that it is a large transition in terms of climate composition across the planet. Hence, the results suggested that climate change is currently encountering a severe impact in all ecosystems globally in a multimodal mean across the world. Consequently, the relative patterns of climate change across the world in terms of z-score anomalies SSP3 from the ESMs suggested that spatial correlation models for climate changes are severe which indeed a clear threat for the global ecosystem.

Naturally, the results of the use of the z-score anomalies SSP3 from the ESMs in stand-clearing disturbance on global ecosystem revealed the estimated percent of an ecoregion presume to be on transition into another ecoregion considering the global temperature variation. Thus, climate change encounters an average of 1.6–2.1°C global temperature rising for the past decades in respect to day and night and hot and cold regions that are correlated with the global distributions of climate change scenarios (*see* Fig. 3.1).

Risk on Global Ecosystem

The results of the broad risk assessment on global ecosystem due to the climate change conducted by the integrated assessment models (IAMs) incorporation with the Intergovernmental Panel on Climate Change (IPCC) suggested that both the species risk axis and the calculated ecosystem disbursement and ecoregion biome transition mode are in a correlated stature. They are the direct constraints for the mechanistic processes that are responsible for climate change since it impacts the global ecosystem's mortality, CO_2-related effects, or disturbance. Subsequently, the worldwide risk for ecosystem considering mechanistic vegetation model based on Phase 6 (CMIP6) for all earth system simulations in global scale revealed that climate change was responsible for a severe loss of ecosystem for the period of

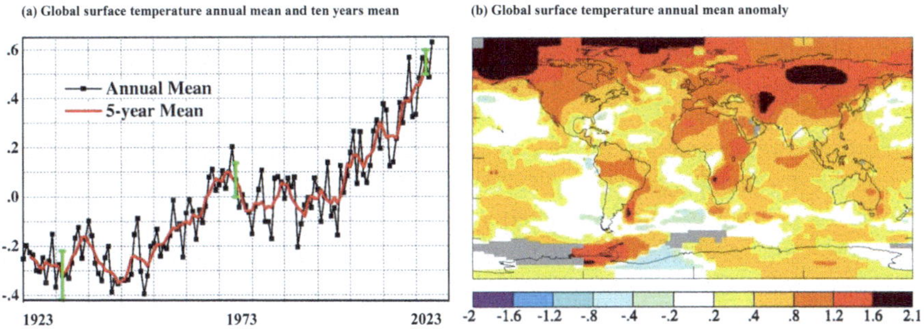

Fig. 3.1 The anticipated average temperature. (**a**) Global surface temperature annual mean and 10 years mean for the period of 1923–2023. (**b**) Global surface temperature annual mean anomaly for the period of 1923–2023

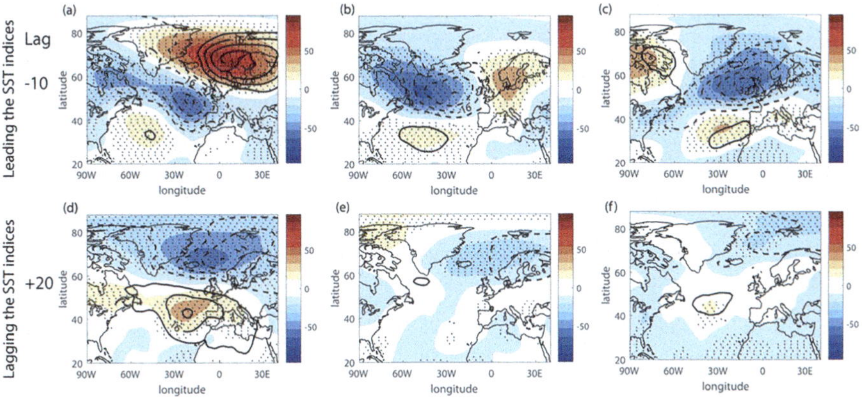

Fig. 3.2 The climate risk projections average on global ecosystem incorporated by the Phase 6 (CMIP6) for earth system models (ESMs) in consideration of SSP3 simulations for the period 1923–2023 in respect to latitude and longitude considering lagging and leading sea surface temperature (SST) indices. (**a**) The multimodel median value considering mean ecosystem considering leading SST indices. (**b**) The range of the change of ecosystem considering leading SST indices. (**c**) The number of ecosystem impact over the number of modes percent ecosystem change globally considering leading SST indices. (**d**) The multimodel median value considering mean ecosystem considering lagging SST indices. (**e**) The range of the change of ecosystem considering lagging SST indices. (**f**) The number of ecosystem impact over the number of modes percent ecosystem change globally considering lagging SST indices

1923–2023 on a global scale in respect to latitude and longitude considering lagging and leading sea surface temperature (SST) indices (*see* Fig. 3.2).

Consequently, the analysis of mechanistic climate change confirms the climate change risk, each of which takes into consideration diverse processes and exerts a significant influence on distinct uncertainties: First, it will effect on global mechanistic ecosystem that has an adverse effect on major climate-sensitive ecosystem distributions, such as dynamic growth disturbances [12, 17]. Second, it has an effect on the characteristics of the ecosystem, such as the biomass, the number of species, or the presence of ecoregion zones. Third, it affects the observational judgment of climate change on ecosystem stand-replacing disturbances, often considering ecosystem globally [1, 11].

Furthermore, to quantify the risk of climate-sensitive disturbances on the global ecosystem for the period of 1923–2023, the two correlative results have been interpreted as follows: (i) a computational random ecosystem model that links the empirical climate change to stand-replacing impacts based on the global scale ecosystem, and (ii) upscaled climate-dependent rates of impacts in the global ecosystem.

The results of both methodical analyses revealed that the estimated impact in stand-replacing the disturbance rates with a climate-driven change severely impact the global ecosystem. Subsequently, a comparison was interpreted between the geographical disturbance of the global ecosystem and the relative risk that was expected to be caused by climate-driven change. Hence, the results suggested that each of the different variables of global ecosystem is significantly related with each other corresponding to the climate change across the world (*see* Fig. 3.2). Indeed, these patterns are robust, in fact, on a global scale, the net damage to the global ecosystem is being sequentially damaged due to climate change which certainly a great risk for the planet earth to sustain in the near future.

Conclusions

The comprehensive and empirical study for the period of 1923–2023 of the CMIP5 data implementation to determine the impact of RCP2.6 on the impacts of biomes suggested that climate change indeed has a great impact on ecosystem vulnerability. Specifically, it changes in species composition throughout the biomes under anomalies in socioeconomic pathway SSP3 from the ESMs and ERA5 in the past recent decades. Subsequently, the integrated assessment models (IAM) and the Intergovernmental Panel on Climate Change (IPCC) suggested that a comprehensive risk on ecosystem in an imminent threat since global ecosystems is imperiled due to the increasing of the carbon pool in the environment. Hence, the result revealed that global ecosystem has a severe negative effect on the ecosystem considering the biomes loss, and species turnover frequencies and imperil function of the vegetation. Consequently, there will be significant dammed to the global ecosystem more since the impact of climate change in the coming year on global vegetation will be more, which will weaken their capacity to sink CO_2. Simply, climate change has a profound impact on biodiversity, leading to more frequent and severe losses of biomes and alterations in species composition, which results in damaging the global ecosystem due to the inability to sink CO_2 to equilibrate the global environment which indeed a great threat for the sustainment of the planet in the near future.

Acknowledgments It is hereby acknowledged that all conclusions, predictions, and findings detailed in this research were independently performed by the author. Moreover, it has been verified that there are no instances of conflicts of interest related to the publication of this research work in a suitable journal or publisher.

References

1. Anderegg, W. R. L., Chao, W., Acil, N., Carvalhais, N., Pugh, T. A. M., Sadler, J. P., & Seidl, R. (2022). A climate risk analysis of Earth's forests in the 21st century. *Science, 377*(6610), 1099–1103.
2. Bauer, J. E., et al. (2013). The changing carbon cycle of the coastal ocean. *Nature, 504*(7478), 61–70.
3. Betts, R. A., et al. (2016). El niño and a record CO_2 rise. *Nature Climate Change, 6*(9), 806–810.
4. Bonan, G. B., et al. (2008). Forests and climate change: Forcings, feedbacks, and the climate benefits of forests. *Science, 320*, 1444–1449.
5. Davis, S. J., et al. (2010). Consumption-based accounting of CO_2 emissions. *Proceedings of the National Academy of Sciences, 107*(12), 5687–5692.
6. de Oliveira Aparecido, L. E., Dutra, A. F., de Lima, R. F., de Alcântara Neto, F., et al. (2022). Climate change scenarios and the dragon fruit climatic zoning in Brazil. *Theoretical and Applied Climatology, 149*, 897–913.
7. Erb, K., et al. (2013). Bias in the attribution of forest carbon sinks. *Nature Climate Change, 3* (10), 854–856.
8. Galina, C. (2013). *An introduction to carbon cycle science*. Cambridge University Press.
9. Hossain, M. F. (2022). Extreme level of CO_2 accumulation into the atmosphere due to the un-equilibrium global carbon emission and sequestration. *Water Air Soil Pollution, 233*, 105.
10. Houghton, R. A. (2007). Balancing the global carbon budget. *Annual Review of Earth and Planetary Sciences, 35*(1), 313–347.
11. Le Quéré, C. (2013). The global carbon budget 1959–2011. *Earth System Science Data, 32*, 44–49.
12. Li, W., et al. (2016). Reducing uncertainties in decadal variability of the global carbon budget with multiple datasets. *Proceedings of the National Academy of Sciences, 113*(46), 13104–13108.
13. Liu, Z., Davis, S. J., Feng, K., Hubacek, K., Liang, S., Anadon, L. D., Chen, B., Liu, J., Yan, J., & Guan, D. (2015a). Targeted opportunities to address the climate–trade dilemma in China. *Nature Climate Change, 6*, 201–206.
14. Liu, Z., et al. (2015b). Reduced carbon emission estimates from fossil fuel combustion and cement production in China. *Nature, 524*(7565), 335–338.
15. Loureiro, I., Ribeiro, C. A., & Sepúlveda, D. (2022). Guimarães 2030: A governance ecosystem. *EuroMediterranean Journal for Environmental Integration, 7*, 319–328.
16. Mason, E. J., et al. (2012). Timing of carbon emissions from global forest clearance. *Nature Climate Change, 2*(9), 682–685.
17. Qiu, L., Jiang, K., Li, Q., Yuan, D., Chen, J., Yang, B., & Achterberg, E. P. (2024). Variability of total alkalinity in coastal surface waters determined using an in-situ analyzer in conjunction with the application of a neural network-based prediction model. *Science of the Total Environment, 908*, 168271.

18. Soengas, P., Rodríguez, V. M., Velasco, P., & Cartea, M. E. (2018). Effect of temperature stress on antioxidant defenses in *Brassica oleracea*. *ACS Omega, 3*(5), 5237–5243.
19. Wang, B., Chen, T., Guobao, X., Guoju, W., & Liu, G. (2023). Management can mitigate drought legacy effects on the growth of a moisture-sensitive conifer tree species. *Forest Ecology and Management, 544*, 121196.

Chapter 4
Remodeling of Photosynthesis to Equilibrate Global CO_2 Cycle

Introduction

There has been a large increase in the concentration of carbon dioxide in the atmosphere due to the rapid acceleration of industrial development around the world [3, 7]. Several studies have shown that the amount of carbon dioxide that is being accumulated in the atmosphere is currently increasing at a rate of 1.418% each year. As a result, the amount of CO_2 that is being accumulated on a global scale is increasing in a rational manner, which poses a significant risk for the next generation to be able to take fresh air to breathe in the near future [2, 12, 17]. Thus, it is the time to ensure that the global climate change condition must be maintained at a consistent and comfortable level by absorbing an excessive amount of CO_2 from the air. This will demonstrate the adaptability, versatility, and manageability of our planet, which will secure a sustainable world for our future generation to live on the earth.

Hence, in this research, a model has been developed to absorb the excessive amount of atmospheric CO_2 naturally by the reformation process of the photosynthesis [13, 21]. Naturally this approach will increase the CO_2 absorption from the air by the process of electric field energy, encoded photosynthesis to sequester excessive amount of CO_2 from the atmosphere to make our planet an environment friendly one.

Methods and Simulation

Modeling of Electric Field Energy (E_e)

A model has been proposed by implementing electric field energy-generated iron stick around the tree in aiming the double up the photosynthesis process. Here, the electrically charged particles will be surrounded by the electric field; as a result,

© The Author(s), under exclusive license to Springer Nature Switzerland AG 2025

M. F. Hossain, *Sustainable Global Development in Environment, Energy, Building, and Infrastructure*, https://doi.org/10.1007/978-3-031-84429-4_4

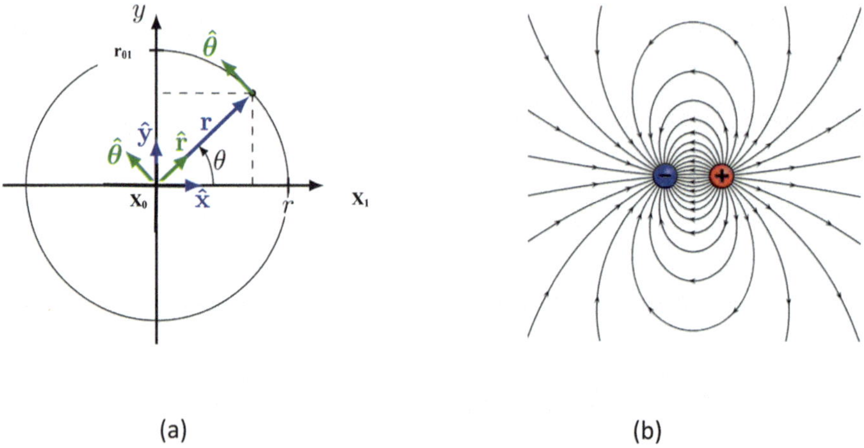

Fig. 4.1 (**a**) Conceptual electric field of a tree, and (**b**) interpretation of the electric field, a tree surrounded by both positive (red) and negative (blue) charges

electrical charge will exert repulsive forces on one another that are negative and positive charges [5, 9]. Therefore, this electric field will consist of one charge or a collection of charges that demonstrate their ability to apply force to another charged force [3, 18, 22]. Hence, Coulomb's Law clarifies that these forces increase with the relation to the distance between these two charges and their magnitudes [4, 6]. Subsequently, the stronger electric field that corresponds to the given distance within the electrostatic force per unit of charge are being applied [8, 11]. Hence, the electric field is being categorized as a function of the electric field; thus, the forces from the electric field have been calculated considering Coulomb's Law where the particles are positioned with the electric charges that can be expressed as follows:

$$F_{01} = \frac{q1\ q0}{4\pi\mathcal{E}_0}\frac{\hat{\boldsymbol{r}}_{01}}{\mathrm{r}_{01}^2} \tag{4.1}$$

where F_{01} is representing the force on charged particle q_0 due to the charged particle q_1; $\mathcal{E}_0$ representing the permittivity of the distance; $\hat{r}_{01}$ denoting a unit vector toward from x_1 to x_0; and r_{01} is the distance from x_1 to x_0 (*see* Fig. 4.1).

Here, $\mathcal{E}_0$ permittivity is substituted for the $\mathcal{E}$ permittivity due to the charges that are being presented in an empty medium. This is due to the fact that the charges q_0 and q_1 exert the same positive force, which are repelled by one another. Subsequently, the particles will have opposite charges, thus, it will attract each other [10, 14]. Therefore, the Coulomb force on any charge at position x_0 is being divided by q_0 considering the other charge by expressing Eq. 4.2:

$$\mathrm{E}_1(x_0) = \mathrm{F}_0 1/q_0 \tag{4.2}$$

where $E_1(x_0)$ is denoting the electric energy field due to the charge q_0 and to q_1. Hence, the electric field energy of a tree at point x_0 is due to the charge q_1 that is a scalar-directed variable equivalent to the Coulomb force per unit charge x_0 [16, 24]. As a result, this force exerted on a charge of rate q at any given position that would be equivalent to the electric field of a tree at any point that can be written as

$$\mathrm{F} = q\mathrm{E} \tag{4.3}$$

Here, the electric field is to be calculated as (N/C), or volt per meter (V/m); considering kg·m·s^{-3}A^{-1}.

Subsequently, the superposition principle will be satisfied by the electric field energy of a tree as the net electric field of a tree is equivalent to the vectorial sum of the electric fields [15, 23]. Thus, the resulting electric field of a tree would be the sum of all electrical fields generated by each particle that can be expressed as

$$\mathrm{E}(\mathrm{x}) = \mathrm{E}_1(\mathrm{x}) + \mathrm{E}_2(\mathrm{x}) + \mathrm{E}_3(\mathrm{x}) + \ldots = \frac{1}{4\pi \mathcal{E}_0} \sum_{k=1}^{N} q_k \frac{\widehat{r}_k}{r_k^2} \tag{4.4}$$

where $\widehat{r}_k$ denotes the vectorial unit to the directional point x_k to point x and r_k represents the distance from x_k to point x.

Here, the electric energy field resulting from the distribution of charge density $\rho(\mathrm{x})$ within the surrounding tree can be determined by considering each unit space dv at point x^1 as a charge $\rho(x^1)dv$. This will allow to calculate the resulting electric field $d\mathrm{E}(\mathrm{x})$ surrounding the tree at point x as

$$d\mathrm{E}(\mathrm{x}) = \frac{\rho\,(x')}{4\pi \mathcal{E}_0} \frac{\widehat{r'}}{r'^2}\, dv \tag{4.5}$$

where the unit vector $\widehat{r'}$ represents the direction from x^1 to x and r', while the direction from x^1 to x and r' represents the distance between x^1 and x. The total field is, therefore, determined by integrating the charge density, which is twice the value of {2E (x)}, over the volume (V), and can be expressed as

$$2\mathrm{E}(\mathrm{x}) = \frac{1}{4\pi \mathcal{E}_0} \iiint_V \rho(x') \frac{\widehat{r'}}{r'^2}\, dv \tag{4.6}$$

As the electric energy charges are two times higher, this electric field energy will pave in taking two-time higher intensity of light from the solar irradiance in respect to the photon frequency (f), and wavelength (λ).

Modeling of Photon Frequency (f) and Wavelength (λ)

Then the charge carrier dynamics of the photon irradiance that is surrounding by a tree are being computed by elucidating the emission rate by the utilization *photoluminescence (PL)* spectroscopy, which corresponds to solar irradiance reaching on the tree. Following that, it was linked to *MATLAB*-generated *Lumiso Pooltest 6 Photometer (LP6)* to connect it with the *SENTINELTM 880 Series* device to clarify the photon irradiance rate on a tree measured by *PerkinElmer® Lambda 25 UV-Vis Spectrometer* precisely [1, 20]. Hence, the primary function of the photon is considered the preservation of solar momentum; thus, all the conserved quantum numbers of solar energy are being maintained calculatively through the control of a *SENTINELTM 880 Series* device which is connected to the *MATLAB* software [19, 26]. After that, the rate of photon irradiance is being calculated in relation to the photon energy trapping by a tree. This is being done by considering the photonic dynamics, which include the basic kinematics, and energy transfer to determine the required photon frequencies (f) and wavelength (λ) to absorb double-efficiency solar energy and pave the way for a process that can perform double the rate of photosynthesis.

Results and Discussion

Modeling of Electric Field Energy (E_e)

Since the electric field energy (E_e) is uniform field that is constant at every point, therefore, the momentum of the electric field energy can be written as

$$E_e = -\frac{\nabla V}{d} \tag{4.7}$$

where ∇V represents the potential variance within the charges, whereas d represents the distance within the charges. The negative function developed here due to the repulsion of positive charges causes a positive charge to be pushed away from the negative charge, thus, the total energy in the electric field per unit volume is being calculated as

$$u_{\mathrm{EM}} = \frac{\varepsilon}{2}|\mathrm{E}|^2 + \frac{1}{2\mu}|\mathrm{B}|^2 \tag{4.8}$$

where ε represents the permittivity of the medium in which the field exists μ, and E and B are the electric field vectors.

Here, the total energy u_{EM} conserved in the electric field in a given volume V can be expressed as

$$u_{\mathrm{EM}} = \frac{1}{2} \int_V \left(\varepsilon \left| \mathrm{E}^2 \right| + \frac{1}{\mu} \left| \mathrm{B}^2 \right| \right) dV \tag{4.9}$$

Hence, this electric field is the primary factor for the vectorial direction, thus, it can be expressed as

$$\mathrm{D} = \varepsilon_0 \mathrm{E} = \mathrm{P} \tag{4.10}$$

Here, P represents the electric polarization, and D represents the electric displacement field. As the E and D fields are correlated to the permittivity of the medium, ε, the electric charges are be proportioned respecting the location and material and it will remain steady that can be expressed as

$$\mathrm{D}(r) = \varepsilon \mathrm{E}(r) \tag{4.11}$$

Hence, the location dependence across the material for inhomogeneous materials, which can be summarized as follows:

$$\mathrm{D}(r) = \varepsilon(r) \mathrm{E}(r) \tag{4.12}$$

As the E and D fields are not parallel in anisotropic materials, the permittivity relates to E and D, the component form can be written as

$$Di = \varepsilon_{ij} \, E_j \tag{4.13}$$

Here, the electric field of point charges moving in a straight line can be found by using the Lorentz transformation of the four forces [4, 22]. Subsequently, the electric field can be confirmed using the form of the Lorentz force into a point charge moving in a straight line that can be expressed as

$$\mathrm{E} = \frac{q}{4\pi\epsilon_0 r^3} \frac{1 - \beta^2}{\left(1 - \beta^2 \sin 2\theta\right)^{3/2} r} \tag{4.14}$$

Here, the variables in the equation are as follows: q represents the electrical charges, r represents the position vector, β represents the ratio of the measured speed of the charged particle, and θ represents the angle between r and the speed of the charge.

To demonstrate the field lines of moving charges along with the charges including the lines would appear to be evenly spaced, the overall electric fields are being computed based on the concept of retarded time, which signifies that electromagnetic fields are related to the speed of light [7, 15]. Thus, the electric field lines form in a continuous path, resulting in the generation of an electromagnetic force of radiation [4, 12]. Simply, this force is related to the boundary of the electric field and the speed of light. To account for the dynamic of the charges, such as Lorenz gauge fields, at the speed of light, the Liénard–Wiechert potential is, thus, used and is expressed as

$$E(r,t) = \frac{1}{4\pi\varepsilon_0}\left(\frac{q(n_s - \beta_s)}{\gamma^2\,(1 - n_s.\beta_s)^2.|r - r_s|^2} + \frac{qn_s x((n_s - \beta_s)x\dot{\beta}_s}{c(1 - n_s.\beta_s)^3|r - r_s|}\right)_{t=t_r} \tag{4.15}$$

Here, the q represents the charge, t_r represents the retarded time, $r_s(t)$ represents the position vector of the particle, $n_s\,(r, t)$ represents a unit vector, $\beta_s(t)$ represents the velocity of the particle, and γ represents the corresponding Lorentz factor, which can be expressed as

$$t_r = t - \frac{|r - r_s(t_r)|}{c} \tag{4.16}$$

Hence, in the case of charged particles that are moving at a speed that is slower than the speed of light, the uniqueness of the solution for t_r for given t, r, and r_s (t) is accurate. Thus, the acceleration-dependent electromagnetic force is produced by accelerating charges considering the advanced time t_a instead of retarded time given as follows:

$$t_a = t + \frac{|r - r_s(t_a)|}{c} \tag{4.17}$$

Hence, the electric field is determined by the particle's condition at a specific moment, it is considered a precise calculation to pave high intensity of light energy intake by a tree to enable much higher photosynthesis process rate in related to the photon frequency (f) and wavelength (λ).

Modeling of Photon Frequency (f) and Wavelength (λ)

Then the charge monitoring is being carried out upon linked to the SENTINEL™ 880 Series electric energy-generated iron rod to determine the identification of the photonic energy release rate due to the electric energy-generated installed iron rod [12, 16]. Consequently, this photon energy release has been computed by measuring the converted particle mass of the photon respecting Einstein's equation, $E = m\,c^2$, where E denotes energy, m denotes mass, and c denotes speed of light [3, 22].

Then the monitoring of the fundamental kinematics of photons is carried out by employing four vector quantities to ascertain the preservation of energy-momentum both before and after the interaction that takes place within the photonic particles. This is accounted for characterizing the formation of two electrons, one of which is negative and the other is positive (positron). Thus, this formation of two electrons is the result of photonic electromagnetic energy traveling through the atomic nucleus, therefore, this photon is an example of the direct conservation of solar energy to matter [4, 7]. Subsequently, the direct conservation of radiant energy is being monitored with the use of a *PerkinElmer*® *Lambda 25 UV-Vis Spectrometer*, and then, the electric energy of the photon is also being monitored to verify that it possesses a minimum mass of two electrons. This is fact that the mass m of a single electron is equivalent to 512,000 electron volts (*eV*) of energy E, which is computed from the equation of Albert Einstein, $E = mc^2$. Therefore, the amount of energy required to produce two electrons should be equal to 1,024,000 *eV* [12, 25]. Then, it will be easily turned into the motion ($E = \left(E_k^{pp}\right)_{tr}$) of the electron–positron, which is being clarified by utilizing a cloud chamber of MATLAB to determine the energy transfer from the photons [7, 19].

Hence, the calculation of the energy transfer to the electron and positron following the generation of the photon is determined as

$$\left(E_k^{pp}\right)_{tr} = hv - 2m_ec^2 \tag{4.18}$$

where h denotes Planck's constant, v denotes the frequency of the photon, and the $2m_ec^2$ denotes the combined rest mass of the electron–positron. As a result of the fact that the electron and the positron are each being generated through the process of pair generation, the average energy transition to each is being computed as follows:

$$\left(E_k^{pp}\right)_{tr} = \frac{1}{2}\left(hv - 2m_ec^2\right) \tag{4.19}$$

By solving these two equations, it can be determined the amount of solar energy needed for a tree by expressing as

$$hv - 2m_ec^2 = \frac{1}{2}\left(hv - 2m_ec^2\right) \tag{4.20}$$

$$hv = 2m_ec^2, \text{ here } hv = E \tag{4.21}$$

$$\text{So}, E = 2mc^2 \tag{4.22}$$

Finally, taking Eq. 4.22 into consideration, a photosynthetic process that is highly efficient can be elucidated by establishing the requisite wavelength (λ) and frequency (f) of the solar irradiation. The transformation of a photon into a particle occurs through the utilization of a gamma ray, which is a result of the reaction between an electron and a positron. As a result, the generation of energy is determined by the famous equation of Einstein $E = mc^2$, where m denotes the mass of the particle, and c denotes the speed of light (3×10^8 ms^{-1}). Considering that the two particles in question are the antiparticles of one another, it follows that their masses are identical and thus, the net energy by the photon can be calculated as follows:

$$E = 2mc^2 \tag{4.23}$$

$$\text{Since } E = hf = \frac{hc}{\lambda} \tag{4.24}$$

Here, the Planck constant, denoted as h, is equal to 6.626×10^{-34} J.s.; the frequency of the photon radiation is denoted as f; subsequently, the speed of light equal to $c = 3 \times 10^8$ ms^{-1} is denoted as c, and the wavelength of photon irradiance is denoted as λ.

Given that the mass of an electron is equal to 9.11×10^{-31} kg and the speed of light is equal to $c = 3 \times 10^8$ ms^{-1}, therefore, it can be concluded that photons are releasing energy in the form of

$$E = 2mc^2 \tag{4.25}$$

$$\text{So, } E = 2^* \left[9.11 \times 10^{-31}\ \text{kg} \times \left(3 \times 10^8\ \text{m/s}\right)^2\right] \tag{4.26}$$

$$= 2^* 8.2 \times 10^{-14}\ \text{j} \tag{4.27}$$

$$= 2^* 8.2 \times 10^{-14}\ \text{j}\ \left(1\text{eV}/1.602 \times 10^{-19}\ \text{j}\right) \tag{4.28}$$

$$= 1{,}024{,}000\ \text{eV} \tag{4.29}$$

For this quantity of energy to be released, a tree must be initiated by the electric energy-generated iron rod to have a plant absorb the following wavelength (λ) and frequencies (f):

$$\lambda = \frac{hc}{2E} \tag{4.30}$$

$$\text{So, } \lambda = \left[\left(6.626 \times 10^{-34}\ \text{J.Sec}\right) \times \left(3 \times 10^8\ \text{m/s}\right)\right] / \left[2 \times \left(8.2 \times 10^{-14}\ \text{j}\right)\right] \tag{4.31}$$

$$\lambda = 1.21 \times 10^{-12}\ \text{m} \tag{4.32}$$

$$\text{and the } E = hf \tag{4.33}$$

$$E = 16.4 \times 10^{-14}\ \text{J} = \left(6.626 \times 10^{-34}\right) \times f \tag{4.34}$$

$$\text{So, } f = 2.5 \times 10^{20}\ \text{Hz} \tag{4.35}$$

Hence, based on the aforementioned calculative results from Eqs. 4.32 and 4.35, it is evident that to achieve a double-rate photosynthesis process in a tree, it is required to have a photonic wavelength (λ) of 1.21×10^{-12} m (Eq. 4.32) and a frequency (f) of 2.5×10^{20} Hz (Eq. 4.35).

Conclusion

The research confirms that global CO_2 cycle can be equilibrated by the process of remodeled photosynthesis considering the generation of electric field energy (D_e). Given that the electric field energy which is exerted by an electric charge on other charges in its array that the force can significantly alter the absorption of light intensity at higher rate. Thus, it will naturally drive the process of photosynthesis at higher rate, which enables a tree to absorb large amounts of carbon dioxide from the air. The results suggested that only photon wavelength (λ) of 1.21×10^{-12} m and frequency (f) of 2.5×10^{20} Hz are required to secure 200% photosynthesis efficiency of a tree upon implementation of an electric field energy releasing iron rod in a tree to absolved CO_2 double rate. Concretely, the application of an electric field energy (D_e) on a tree would make it possible for the acceleration of photosynthetic process to sequester CO_2 from the atmosphere, which indeed be an excellent finding among the scientist communities in contributing to mitigate the climate change crisis dramatically.

Acknowledgments This research does not have any conflicts of interest that would prevent it from being published by an appropriate publisher.

References

1. Artemyev, N., Jentschura, U., Serbo, V., & Surzhykov, A. (2012). Strong electromagnetic field effects in ultra-relativistic heavy-ion collisions. *European Physical Journal C: Particles and Fields, 72*, 1935.
2. Birnbaum, K. M., et al. (2005). Photon blockade in an optical cavity with one trapped atom. *Nature, 436*, 87–90.
3. Busch, K., et al. (2007). Periodic nanostructures for photonics. *Physics Reports, 444*, 101–202.
4. Cartwright, J. (2010). Photons meet with three-way split. *Nature*. https://doi.org/10.1038/news.2010.381
5. Chang, D. E., Sørensen, A. S., Demler, E. A., & Lukin, M. D. (2007). A single-photon transistor using nanoscale surface plasmons. *Nature Physics, 3*, 807–812.
6. Chen, G., et al. (2013). Examining non-locality and quantum coherent dynamics induced by a common reservoir. *Scientific Reports*.
7. Dayan, B., et al. (2008). A photon turnstile dynamically regulated by one atom. *Science, 319*, 1062–1065.
8. Douglas, J. S., et al. (2015). Quantum many-body models with cold atoms coupled to photonic crystals. *Nature Photonics, 9*, 326–331.
9. Eichler, J., & Stöhlker, T. (2007). Radiative electron capture in relativistic ion–atom collisions and the photoelectric effect in hydrogen-like high-Z systems. *Physics Reports, 439*, 1–99.
10. Englund, D., et al. (2010). Resonant excitation of a quantum dot strongly coupled to a photonic crystal nanocavity. *Physical Review Letters, 104*, 073904.
11. Guerlin, C., et al. (2007). Progressive field-state collapse and quantum non-demolition photon counting. *Nature, 448*, 889–893.
12. Javadi, A., et al. (2015). Single-photon non-linear optics with a quantum dot in a waveguide. *Nature Communications, 6*, 8655.

13. Jiaqi, L., Yi, Z., Chengkai, T., & Zhu, X. (2019). *INS aided high dynamic single-satellite position algorithm.* In 2019 IEEE international conference on signal processing, communications and computing (ICSPCC).
14. Kaneda, F., et al. (2019). High-efficiency single-photon generation via large-scale active time multiplexing. *Science Advances, 5*(10), eaaw8586.
15. Langer, L., et al. (2014). Access to long-term optical memories using photon echoes retrieved from semiconductor spins. *Nature Photonics, 8*, 851–857.
16. Langford, K., et al. (2021). Efficient quantum computing using coherent photon conversion. *Nature, 478*, 360–363.
17. Lodahl, P., Mahmoodian, S., Stobbe, S., Rauschenbeutel, A., et al. (2017). Chiral quantum optics. *Nature.*
18. Lü, X.-Y., Zhang, W.-M., Ashhab, S., Ying, W., & Nori, F. (2013). Quantum-criticality induced strong Kerr nonlinearities in optomechanical systems. *Scientific Reports.*
19. Naghiloo, M., et al. (2016). Mapping quantum state dynamics in spontaneous emission. *Nature Communications.*
20. Najjari, B., Voitkiv, A. B., Artemyev, A., & Surzhykov, A. (2009). Simultaneous electron capture and bound-free pair production in relativistic collisions of heavy nuclei with atoms. *Physical Review A, 80*, 012701.
21. Prasad, A. S., Hinney, J., Mahmoodian, S., Hammerer, K., et al. (2020). Correlating photons using the collective nonlinear response of atoms weakly coupled to an optical mode. *Nature Photonics.*
22. Reinhard, A., et al. (2012). Strongly correlated photons on a chip. *Nature Photonics, 6*, 93–96.
23. Shalm, L. K., et al. (2012). Three-photon energy–time entanglement. *Nature Physics.*
24. Ting, T. C. (2004). The polarization vector and secular equation for surface waves in an anisotropic elastic half-space. *International Journal of Solids and Structures.*
25. Vanner, M. R., Hofer, J., Cole, G. D., & Aspelmeyer, M. (2013). Cooling-by-measurement and mechanical state tomography via pulsed optomechanics. *Nature Communications.*
26. Wang, C., Wang, W., & Chen, Z. (2017). *Single-satellite positioning algorithm based on direction-finding.* In 2017 Progress in electromagnetics research symposium - Spring (PIERS).

Chapter 5
Application of Dormant Light Energy to Breakdown CO_2 Naturally

Introduction

Mismanagement of the world's natural resources is hastening the depletion of CO_2, which is a major cause of climate change [2, 7, 42]. Multiple studies have shown that CO_2 emissions are currently climbing into the atmosphere at a pace of 1.418% each year, leading to more severe mankind life on the earth increasingly precarious for generations to come [3, 11, 36]. It is definitely time to equilibrate the atmospheric CO_2 to make the world climate stable and comfortable condition for the future generation. This will prove that our mother earth is versatile, adaptable, and manageable; doing so will not cause maladjustment but will leave a better planet for the next generation to enjoy. So, there has been effort in this study to naturally break down CO_2 to neutralize the climate change by utilizing the activation of the latent particle of light energy that is present all over the planet [6, 21, 35]. Hence, this innovative mathematical model of the earth's latent light energy's activation to decompose atmospheric CO_2 will undoubtedly be an excellent scientific discovery in the quest to naturally control the climate change.

Methods and Simulation

Modeling of Dormant Solar Energy ($\mathbf{D_e}$)

To better understand this latent energy on the earth, the author computed its dormant light energy taking the sphericity and orbital parameters into account and expressed it as follows:

© The Author(s), under exclusive license to Springer Nature Switzerland AG 2025
M. F. Hossain, *Sustainable Global Development in Environment, Energy, Building, and Infrastructure*, https://doi.org/10.1007/978-3-031-84429-4_5

$$\rho\cos(c) = \rho\cos(a)\cos(b) + \rho\sin(a)\sin(b)\cos(C) \quad (5.1)$$

Here, C represents the vertex angle that corresponds to the arc of the earth considering length c, and a, b, c are the radians that correspond to the earth spherical triangle [8, 24, 38]. It is obvious that the terms that follow may be stated as the laws of cosines when the zenith angle Θ is taken into account, and thus, the equation can, therefore, be written as

$$\rho\cos(\Theta) = \sin(\phi)\sin(\delta) + \cos(\phi)\cos(\delta)\cos(r) \quad (5.2)$$

In this context, ϕ represents the horizontal angle of the earth, δ represents the azimuth angle, and ρ represents the earth's sphere. r is the mean distance corresponding to astronomical unit, and S_0 represents the dormant light energy [12, 14, 18]. So, the amount of latent light energy that falls onto the earth's tangent plane is calculated by

$$Q = \begin{cases} S_o \frac{R_0^2}{R_E^2} \cos(\theta) & \cos(\theta) > 0 \\ 0 & \cos(\theta) \leq 0 \end{cases} \quad (5.3)$$

Here, the average Q^{-} over a time is the average value of Q corresponding to its rotation of h = π to h = −π, and thus, equation is being expressed as

$$Q^{-time} = -\frac{1}{2\pi}\int_{\pi}^{-\pi} Q dh \quad (5.4)$$

During the latent light energy reaching on the earth, then, Θ = 1 /2 π, or for h_0; therefore, the equation can be clarified by the time-related angle as

$$\rho\sin(\phi)\sin(\delta) + \cos(\phi)\cos(\delta)\cos(h_0) = 0$$

or

$$\rho\cos(h_0) = -\tan(\phi)\tan(\delta) \quad (5.5)$$

Hence, h_o = π, as the latent light energy will affirm as tan(φ)tan(δ) greater than 1. Consequently, in a situation where Q time = 0, it would stay relatively constant throughout the earth's period of inactive light energy generation, which can be described by using following equation:

$$\int_{\pi}^{-\pi} Qdh = \int_{h_0}^{-h_0} Qdh = S_0 \frac{R_0^2}{R_E^2} \int_{h_0}^{-h_0} \cos(\theta) dh = S_0 \frac{R_0^2}{R_E^2}$$
$$\times \left[h\sin(\phi)\sin(\delta) + \cos(\phi)\cos(\delta)\sin(h)\right]_{h=h_0}^{h=-h_0} = -2S_0 \frac{R_0^2}{R_E^2}$$
$$\times \left[h_0\sin(\phi)\sin(\delta) + \cos(\phi)\cos(\delta)\sin\left(h_{0)}\right] \tag{5.6}$$

Therefore

$$Q^{-time} = \frac{S_0}{\pi} \frac{R_0^2}{R_E^2} \left[h_0\sin(\phi)\sin(\delta) + \cos(\phi)\cos(\delta)\sin\left(h_{0)}\right] \tag{5.7}$$

Here, θ is representing horizontal angle that corresponds to the orbit of the earth, thus, θ = 0 corresponds to its vernal equinox of δ at its orbital position that can be expressed as

$$\delta = \varepsilon \sin(\theta) \tag{5.8}$$

ε denotes the horizontal longitude of sphere ϖ in relation to vernal equinox, and therefore, the equation can be rewritten as

$$R_E = \frac{R_0}{1 + e\cos(\theta - \omega)} \tag{5.9}$$

or

$$\frac{R_0}{R_E} = 1 + e\cos(\theta - \omega) \tag{5.10}$$

Subsequently, the latent energy can be clarified from any latitude φ and θ of the earth, based on the astro-dynamical computation of ϖ, ε, so, and e [13, 37, 39]. To simplify the equation for latent light energy generation at a given moment on the earth, the author used the values of θ = 0° for the specific time, and thus, θ = ∞ for the precise horizon of the equinox.

$$D_e = S_0\left(1 + \cos\left(2\pi\frac{n}{x}\right)\right) \tag{5.11}$$

Here, n represents unit of time, x represents the period of time, and z represents the spherical earth that maximizes the formation of latent light energy, which corresponds to the hidden particles on the earth in relation to law of cosines and the sphericity of the earth (*see* Fig. 5.1).

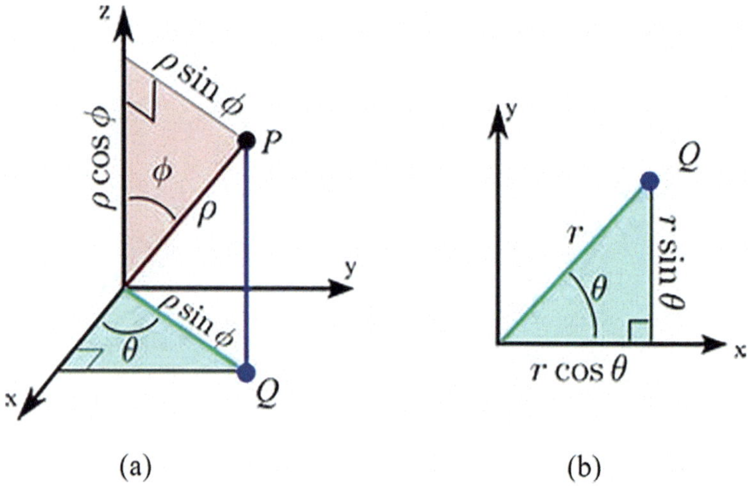

Fig. 5.1 The dispersion of earth's latent light energy: (**a**) an explanation of its existence in relation to the law of cosines, and (**b**) a representation, schematically, of the dynamical effects of latent light energy on the sphericity of our planet

Activation of Dormant Light Energy (D_e) to Break the CO_2

Owing to the fact that latent light energy is similar to active light energy once the particle of dormant energy is being excited due to the active U(1) gauge field in terms of the vectorial (V) dynamics of its tangential (T) plane of the earth [1, 9, 10]. As a result, the dormant light energy will be initiated as a kinetic energy corresponding to the dormant solar energy field of the earth as

$$\overrightarrow{De} = \left|\frac{Edp}{c}, \overrightarrow{Pd}\right| + \overrightarrow{Xef}\left[\frac{Eef}{c}, \overrightarrow{Pef}\right] \tag{5.12}$$

The equation can be restated as follows: as the momentum of latent solar energy is zero and it is conserved with its hypercharge electron, the sum of the beginning momenta is equivalent to the sum of the final momentum that can be expressed as

$$\overrightarrow{De} + \overrightarrow{Xei} = \overrightarrow{Xp} + \overrightarrow{Xef} \tag{5.13}$$

Subsequently, I have rewritten the equation by taking the square of both sides, which represents the computed isolated momentum of the exiting hypercharge electron, as

$$\left(\overrightarrow{De} + \overrightarrow{Xei} - \overrightarrow{Xp}\right)^2 = \overrightarrow{Xef}^2 \tag{5.14}$$

By adding the cross term to the vectors of this latent solar energy's hypercharge, electron momentum can be expressed as

$$\overrightarrow{De} = 2\,\overrightarrow{Xd}.\overrightarrow{Xei} = 2 * \left[\frac{Ed}{c}, \overrightarrow{Pd}\right].\left[m_e c, \vec{0}\right] = 2Edm_e \tag{5.15}$$

Then, to represent hypercharge electron creation, I can multiply all resultant forces of the initially stationed electron by the generation of latent light energy, and I can get the following equation:

$$\overrightarrow{De} = m_d^2c^2 - m_e^2c^2 + -2m_eE_p - 2\left(\frac{E_dE_p}{c^2} - \frac{P_dE_p\cos(\theta)}{c}\right) + 2E_dm_e \tag{5.16}$$

which can be simplified as

$$\overrightarrow{De} = 2\left(m_eE_p + \left(\frac{E_dE_p}{c^2}\right) - \frac{P_dE_p}{c}\cos(\theta)\right) = m_d^2c^2 - 2E_dm_e \tag{5.17}$$

In this context, the relation between the scalar momentum of latent light energy and the hypercharge electron generation expression yields that $m_d \rightarrow 0$, which can be stated in terms of scattering [15, 22, 41]. Here, the latent light energy is a striking electron force and can be described by the hypercharge electron creation as $m_d \rightarrow 0$, and $P_d \rightarrow Ep \rightarrow E_d$.

Thus, the enormous active force of this striking electron force of latent light energy will disperse the electromagnetic wave at a speed of light c [4, 25, 32]. This tremendous momentum of the hypercharge-hitting electron force will be enough to break the CO_2 covalent bond (C=O) and rapidly will convert it into carbon and oxygen because this force is sensitive to this covalent bond [16, 40, 41]. Because of the high degree of kinematicity in this function of the hypercharge impacting on the electron of the latent light energy, Eq. 5.18 can be described as

$$\overrightarrow{De} = 2\left[\frac{Edp}{c}, \overrightarrow{Pd}\right] + \overrightarrow{CO_2} + \left[\overrightarrow{Pemf}\right] \tag{5.18}$$

$$= 2\left[\frac{Edp}{c}, \overrightarrow{Pd}\right] + \left[\overrightarrow{C = O_2}\right] - [\text{Covalent Bond}] \tag{5.19}$$

As dormant light energy instigation is constant, force of the outgoing electron of the dormant light energy can be rewritten as

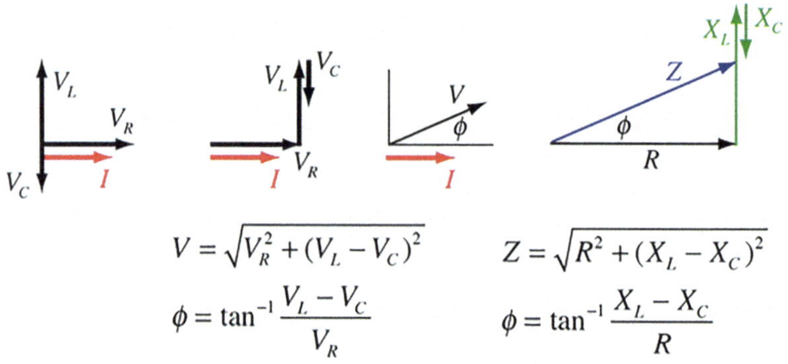

Fig. 5.2 The base mode of dormant light energy U(1) gauge asymmetry on the earth related to its scalar dynamics

$$\overrightarrow{De}=2\left[\frac{Ed}{c},\overrightarrow{Pd}\right]+\left[\overrightarrow{C+O2}\right]-\left[m_e\overrightarrow{c^2}\right]=2Edm_e \tag{5.20}$$

As $\overrightarrow{De}=S_0\left(1+\cos\left(2\pi\frac{n}{x}\right)\right)$ and CO_2 can be spitted rapidly by this hypercharged electron of dormant light energy force, the equation can be rewritten as (Fig. 5.2)

$$\overrightarrow{De}=2E_dm_e=S_0\left(1+\cos\left(2\pi\frac{n}{x}\right)\right)-[C+O_2] \tag{5.21}$$

Results and Discussion

Modeling of Dormant Light Energy (D_e)

The modeling of dormant light energy is being clarified by the following equation which is actually the latent light energy irradiance released from the Earth's gauge field U(1) at a scalar dynamic of its tangential direction [17, 19, 34]. Therefore, the equation can be expressed as

$$D_e=P_{pv}=\eta_{pvg}\,A_{pvg}\,G_t \tag{5.22}$$

Here, ηpc represents the efficiency of dormant light energy formation, β is the time cofactor, ηr represents the mode of dormant light energy efficiency, and T_{cref} represents the tangential condition. Therefore, the equation can be rewritten as

$$\eta_{pvg} = \eta_r \eta_{pc}\left[1 - \beta\left(T_c - T_{c\ ref}\right)\right] \tag{5.23}$$

Here, η_{pc} denotes the dormant light energy, β represents the time-related factor, η_r represents the mode of dormant light energy, and T_{cref} represents the condition of tangential; therefore, the formula can be expressed by

$$T_c = T_a + \left(\frac{NOCT - 20}{800}\right) G_t \tag{5.24}$$

With this assumption of the total earth plane, I can determine the net dormant light radiation on earth's surface using the following equation, where Ta is tangential, and Gt is the energy irradiance on the earth sphere (W/m^2).

$$I_t = I_b R_b + I_d R_d + (I_b + I_d) R_r \tag{5.25}$$

The following rewritten equation, which is based on the link between electron charge and energy, has been calculated for the Earth's surface latent solar energy since dormant energy is inherently acting as a hypothetical superconductor.

$$I = N_p\left[I_{ph} - I_s\left[\exp\left(\frac{q(V + IR_s)}{AKTN_s} - 1\right)\right]\right] \tag{5.26}$$

where

$$I_{rs} = I_{rr}\left(\frac{T}{T_r}\right)^3 exp\left[\frac{E_G}{AK}\left(\frac{1}{T_r} - \frac{1}{T}\right)\right] \tag{5.27}$$

Because of this, in Eqs. 5.5 and 5.6, q represents the electron charge, K represents the Boltzmann's constant, A represents the diode's standardized efficiency, and T is the tangential [5, 11, 23]. So, I_{rs} is the reversal of the Earth's dormant light energy motion at T, I_{rr} is the inverse of this energy at T_r, and EG is the dormant solar energy efficiency of the earth's surface. Accordingly, the expression for the dormant light energy I_{ph} generation on the earth's surface can be expressed as

$$I_{ph} = \left[I_{SCR} + k_i(T - T_r)\frac{S}{100}\right] \tag{5.28}$$

Here, I_{SCR} stands for the energy-motion factor that takes into account the earth's optimal energy radiation dynamic, k_i stands for the energy-momentum factor, and S stands for the calculation of dormant light energy radiation in mW/cm^2 [27, 28, 33]. Consequently, the mathematical model of the net dormant light energy formation can be stated as

$$I = I_{ph} - I_D \tag{5.29}$$

$$I = I_{ph} - I_0 \left[\exp\left(\frac{q(V + R_s I)}{AKT} - 1 \right) \right] \tag{5.30}$$

The following symbols are used in this context, I_{ph} represents dormant light energy-dynamic, I_D represents originated dormant energy dynamics, I_0 represents inversed energy-dynamic, q represents electron charge (1.6 × 10–19 C), K represents Boltzmann's constant, T represents earth tangential, R_s represents energy cofactor, R_{sh} represents energy coefficient, I denotes dormant light energy-motion, and V denotes energy velocity. Thus, the following equation can be used to find the net amount of latent light energy that enters the earth's surface:

$$I = I_{PV} - I_{D1} - \left(\frac{V + IR_S}{R_{SH}} \right) \tag{5.31}$$

where

$$I_{D1} = I_{01} \left[\exp\left(\frac{V + IR_s}{a_1 V_{T1}} \right) - 1 \right] \tag{5.32}$$

The conceptual latent light energy vector is represented by I and I_{01}, the reversed flow of dormant light energy, and the ideal energy state is represented by VT1; thus, the net latent light energy generation on any Earth array can be expressed as

$$v_{oc} = \frac{V_{oc}}{cK\,T/q} \tag{5.33}$$

$$P_{max} = \frac{\frac{V_{oc}}{cK\,T/q} - \ln\left(\frac{V_{oc}}{cK\,T/q} + 0.72 \right)}{\left(1 + \frac{V_{oc}}{K\,T/q} \right)} \left(1 - \frac{V_{oc}}{\frac{V_{oc}}{I_{SC}}} \right) \left(\frac{V_{oc0}}{1 + \beta \ln \frac{G_0}{G}} \right) \left(\frac{T_o}{T} \right)^{y} I_{sc0} \left(\frac{G}{G_o} \right)^{a} \tag{5.34}$$

Here, the following symbols are used in this context, where v_{oc} represents flow of energy, Voc represents energy condition at Vt = nkT/q, c denotes constant motion of energy, K is Boltzmann's constant, T represents tangential of earth plane, α represents functional motion of energy, q represents electron charge, γ represents function acting on all nonlinear energy, and β represents earth plane for enhancing energy flowing rate.

Activation of Dormant Light Energy (D_e) to Break the CO_2

To verify the functional test of the latent light energy on CO_2, the analysis of the global Albanian symmetries of the gauge field was carried out to activate the dormant light energy and break the CO_2 on the earth. So, the net dormant light energy particle at the global symmetrical array of the earth sphere has been calculated by the analysis of the dormant light energy, which was initiated by the gauge field of the earth [30, 31, 39]. According to the findings, the earth's spherical complex vector field couples with the electromagnetic field do allow the dormant light energy charge q to be translated into the following equation:

$$P_{max} = \frac{\frac{V_{oc}}{cK\,T/q} - \ln\left(\frac{V_{oc}}{cK\,T/q} + 0.72\right)}{\left(1 + \frac{V_{oc}}{K\,T/q}\right)}\left(1 - \frac{V_{OC}}{\frac{V_{oc}}{I_{SC}}}\right)\left(\frac{V_{oc0}}{1 + \beta \ln \frac{G_0}{G}}\right)\left(\frac{T_o}{T}\right)^y I_{sc0}\left(\frac{G}{G_o}\right)^a \tag{5.35}$$

$$\mathfrak{h} = -\frac{1}{4}\, F_{\mu\nu}F^{\mu\nu} + D_\mu\Phi^*\, D^\mu\Phi - V(\Phi^*\Phi) \tag{5.36}$$

where

$$D_\mu\Phi(x) = \partial_\mu\Phi(x) + \mathrm{iqA_\mu(x)}\Phi\,(\mathrm{x}) \tag{5.37}$$

$$D_\mu\Phi^*(x) = \partial_\mu\Phi^*(\mathrm{x}) - \mathrm{iqA_\mu(x)}\Phi^*(\mathrm{x}) \tag{5.38}$$

and

$$V(\Phi^*\Phi) = \frac{\lambda}{2}\,(\Phi^*\Phi)^2 + m^2(\Phi^*\Phi) \tag{5.39}$$

and

$$v = \sqrt{\frac{-2m^2}{\lambda}}, \text{any real } \theta \tag{5.40}$$

Afterward, the U(1) symmetrical net inactive light energy generation is being determined simultaneously by the non-zero functional value of the global earth sphere's vector field. After verifying the x-dependent condition of the symmetrical (x) array of the earth sphere, the expected value of the global U(1) net symmetrical dormant light energy of (x) can be stated by the following equation:

$$\Phi\,(\mathrm{x}) = \frac{1}{\sqrt{2}}\,\Phi_\mathrm{r}(\mathrm{x}) * \mathrm{e}^{\mathrm{i}\Theta(\mathrm{x})}, \text{real } \Phi_\mathrm{r}(\mathrm{x}) > 0, \text{real } \Phi(\mathrm{x}) \tag{5.41}$$

Next, I take into account the maximum level of latent light energy emission on Earth's sphere by computing the net amount of hypercharge energy creation from the vector on the earth plane [20, 26, 29]. This function is of the first order. Based on the results of the calculation of latent light energy and its vector on the earth field, the following equation confirms the net hypercharge energy generation as

$$V(\phi) = \frac{\lambda}{8}\left(\phi_r^2 - v^2\right)^2 + \text{const}, \tag{5.42}$$

or the resultant hypercharge energy generation is shifted by its VEV, $\Phi_r(x) = v + \sigma(x)$,

$$\phi_r^2 - v^2 = (v+\sigma)^2 - v^2 = 2v\sigma + \sigma^2 \tag{5.43}$$

$$V = \frac{\lambda}{8}\left(2v\sigma - \sigma^2\right)^2 = \frac{\lambda v^2}{2} * \sigma^2 + \frac{\lambda v}{2} * \sigma^3 + \frac{\lambda}{8} * \sigma^4 \tag{5.44}$$

Simultaneously, the functional derivative $D_\mu \phi$ will become

$$D_\mu \phi = \frac{1}{\sqrt{2}}\left(\partial_\mu\left(\phi_r e^{i\Theta}\right) + iqA_\mu * \phi_r e^{i\Theta}\right) = \frac{e^{i\Theta}}{\sqrt{2}}\left(\partial_\mu \phi_r + \phi_r * i\partial_\mu \Theta + \phi_r * iqA_\mu\right) \tag{5.45}$$

$$\begin{aligned} \left|D_\mu \phi\right|^2 &= \frac{1}{2}\left|\partial_\mu \phi_r + \phi_r * i\partial_\mu \Theta + \phi_r * iqA_\mu\right|^2 = \frac{1}{2}\left(\partial_\mu \phi_r\right) + \frac{\phi_r^2}{2} \\ &* \left(\partial_\mu \Theta q A_\mu\right)^2 = \frac{1}{2}\left(\partial_\mu \sigma\right)^2 + \frac{(v+\sigma)^2}{2} * \left(\partial_\mu \Theta + qA_\mu\right)^2 \end{aligned} \tag{5.46}$$

Altogether,

$$\mathfrak{h} = \frac{1}{2}\left(\partial_\mu \sigma\right)^2 - v(\sigma) - \frac{1}{4} F_{\mu\upsilon} F^{\mu\upsilon} + \frac{(v+\sigma)^2}{2} * \left(\partial_\mu \Theta + qA_\mu\right)^2 \tag{5.47}$$

and

$$\mathfrak{h}_{sef} = \frac{1}{2}\left(\partial_\mu \sigma\right)^2 - \frac{\lambda v^2}{2} * \sigma^2 - \frac{1}{4} F_{\mu\upsilon} F^{\mu\upsilon} + \frac{v^2}{2} * \left(qA_\mu + \partial_\mu \Theta\right)^2 \tag{5.48}$$

Hence, the net hypercharge energy generation is known to be a realistic mass-two vector particle which is very much sensitive to carbon dioxide, and thus, the equation may be rewritten as

$$\mathfrak{h}_{sef} = \frac{1}{2}\left(\partial_\mu \sigma\right)^2 - \frac{\lambda v^2}{2} * \sigma^2 - \frac{1}{4} F_{\mu\nu}F^{\mu\nu} + \frac{v^2}{2} * \left(qA_\mu + \partial_\mu \Theta\right)^2$$
$$= 2\left[\frac{Ed}{c}, \overrightarrow{Pd}\right] + \left[\overrightarrow{CO2}\right] - \left[m_e \overrightarrow{c^2}\right] \tag{5.49}$$

So,

$$\mathfrak{h}_{sef} = 2\left[\frac{Ed}{c}, \overrightarrow{Pd}\right] + \left[\overrightarrow{CO2}\right] - \left[m_e \overrightarrow{c^2}\right] \tag{5.50}$$

Here, the hypercharge energy of dormant light energy releases the net momentum on the earth, and thus, the equation can be rewritten as

$$\mathfrak{h}_{sef} = 2\left[\frac{Ed}{c}, \overrightarrow{Pd}\right] + \left[\overrightarrow{C + O2}\right] - \left[m_e \overrightarrow{c^2}\right] \tag{5.51}$$

As $\mathfrak{h}_{sef} = De$, dormant light energy finally can be written as

$$D_e = 2\left[\frac{Ed}{c}, \overrightarrow{Pd}\right] + \left[\overrightarrow{C + O2}\right] - \left[m_e \overrightarrow{c^2}\right] \tag{5.52}$$

Simply, the result suggested that dormant light energy can play a vital role to breakdown CO_2 by the activation of its hypercharge energy momentum within its energy generation fields (x) that address to mitigate the global CO_2 on the earth.

Conclusions

The activation of latent light energy of the earth has been modeled to verify that it has sufficient dormant light energy within the earth to break down atmospheric CO_2. Thus, a number of mathematical tests have performed which suggested that dormant light energy is readily available on the earth that can play a crucial role in breaking down atmospheric CO_2 naturally to address the complexity of climate change. To put it simply, the quantum electrodynamic (QED) momentum of the activated dormant light energy is being modeled to efficiently break down the atmospheric CO_2 into C and O_2 and equilibrate the global carbon cycle in the atmospheric CO_2 to mitigate the climate change. Therefore, this mathematical research will be the most groundbreaking scientific finding of all time undoubtedly to combat the climate change catastrophe dramatically.

Acknowledgments The author asserts that this work does not have any conflict of interest for publication with any appropriate publisher.

References

1. Artemyev, N., Jentschura, U., Serbo, V., & Surzhykov, A. (2012). Strong electromagnetic field effects in ultra-relativistic heavy-ion collisions. *European Physical Journal C: Particles and Fields, 72*, 1935.
2. Baur, G., Hencken, K., Trautmann, D., Sadovsky, S., & Kharlov, Y. (2002). Dense laser-driven electron sheets as relativistic mirrors for coherent production of brilliant X-ray and γ-ray beams. *Physics Reports, 364*, 359–450.
3. Baur, G., Hencken, K., & Trautmann, D. (2007). Revisiting unitarity corrections for electromagnetic processes in collisions of relativistic nuclei. *Physics Reports, 453*, 1–27.
4. Birnbaum, K. M., et al. (2005). Photon blockade in an optical cavity with one trapped atom. *Nature, 436*, 87–90.
5. Boettcher, I., Pawlowski, J. M., & Diehl, S. (2012). Ultracold atoms and the functional renormalization group. *Nuclear Physics B - Proceedings Supplements, 228*, 63–135.
6. Broz, M., et al. (2020). A generator of forward neutrons for ultra-peripheral collisions. *Computer Physics Communications*.
7. Busch, K., et al. (2007). Periodic nanostructures for photonics. *Physics Reports, 444*, 101–202.
8. Cardoso, V., Lemos, J. P., & Yoshida, S. (2004). Quasinormal modes of Schwarzschild black holes in four and higher dimensions. *Physical Review D, 69*, 044004.
9. Cartwright, J. (2010). Photons meet with three-way split. *Nature*. https://doi.org/10.1038/news.2010.381
10. Chang, D. E., Sørensen, A. S., Demler, E. A., & Lukin, M. D. (2007). A single-photon transistor using nanoscale surface plasmons. *Nature Physics, 3*, 807–812.
11. Chen, J., Wang, C., Zhang, R., & Xiao, J. (2012). Multiple plasmon-induced transparencies in coupled-resonator systems. *Optics Letters, 37*, 5133–5135.
12. Cheng, M. T., & Song, Y. Y. (2012). Fano resonance analysis in a pair of semiconductor quantum dots coupling to a metal nanowire. *Optics Letters, 37*, 978–980.
13. Dayan, B., et al. (2008). A photon turnstile dynamically regulated by one atom. *Science, 319*, 1062–1065.
14. Destrade, M. (2003). Elastic interface acoustic waves in twinned crystals. *International Journal of Solids and Structures*.
15. Dobrynina, A., Kartavtsev, A., & Raffelt, G. (2015). Photon-photon dispersion of TeV gamma rays and its role for photon-ALP conversion. *Physical Review D, 91*, 083003.
16. Douglas, J. S., et al. (2015). Quantum many-body models with cold atoms coupled to photonic crystals. *Nature Photonics, 9*, 326–331.
17. Dupuis, N. L., et al. (2021). The nonperturbative functional renormalization group and its applications. *Physics Reports*.
18. Eichler, J., & Stöhlker, T. (2007). Radiative electron capture in relativistic ion–atom collisions and the photoelectric effect in hydrogen-like high-Z systems. *Physics Reports, 439*, 1–99.
19. Englund, D., et al. (2010). Resonant excitation of a quantum dot strongly coupled to a photonic crystal nanocavity. *Physical Review Letters, 104*, 073904.
20. Fernandez, J., & Martín, F. (2009). Electron and ion angular distributions in resonant dissociative photoionization of H2 and D2 using linearly polarized light. *New Journal of Physics, 11*, 043020.
21. Gabor, M., et al. (2009). Extremely efficient multiple electron-hole pair generation in carbon nanotube photodiodes. *Science, 325*, 1367–1371.
22. Gleyzes, S., et al. (2007). Quantum jumps of light recording the birth and death of a photon in a cavity. *Nature, 446*, 297–300.
23. Guerlin, C., et al. (2007). Progressive field-state collapse and quantum non-demolition photon counting. *Nature, 448*, 889–893.
24. Guo, Y., Al-Jubainawi, A., & Ma, Z. (2019). Performance investigation and optimisation of electrodialysis regeneration for LiCl liquid desiccant cooling systems. *Applied Thermal Engineering, 149*, 1023–1034.

25. Hencken, K., Baur, G., & Trautmann, D. (2006). Transverse momentum distribution of vector mesons produced in ultraperipheral relativistic heavy ion collisions. *Physical Review Letters, 96*, 012303.
26. Hübel, H., et al. (2010). Direct generation of photon triplets using cascaded photon-pair sources. *Nature, 466*, 601–603.
27. Javadi, A., et al. (2015). Single-photon non-linear optics with a quantum dot in a waveguide. *Nature Communications, 6*, 8655.
28. Johnson, B. R. "Quantum non-demolition detection of single microwave photons in a circuit", Nature Physics, 06/20/2010.
29. Kevin, A., et al. (2017). Signatures of two-photon pulses from a quantum two-level system. *Nature Physics, 3*, 649–654.
30. Langer, L., et al. (2014). Access to long-term optical memories using photon echoes retrieved from semiconductor spins. *Nature Photonics, 8*, 851–857.
31. Langford, K., et al. (2021). Efficient quantum computing using coherent photon conversion. *Nature, 478*, 360–363.
32. Liu, K., Hong, X., Zhou, Q., Jin, C., Li, J., Zhou, W., Liu, J., Wang, E., Zettl, A., & Wang, F. (2013). High-throughput optical imaging and spectroscopy of individual carbon nanotubes in devices. *Nature Nanotechnology*.
33. Naghiloo, M., et al. (2016). Mapping quantum state dynamics in spontaneous emission. *Nature Communications*.
34. Najjari, B., Voitkiv, A. B., Artemyev, A., & Surzhykov, A. (2009). Simultaneous electron capture and bound-free pair production in relativistic collisions of heavy nuclei with atoms. *Physical Review A, 80*, 012701.
35. Prasad, A. S., Hinney, J., Mahmoodian, S., Hammerer, K., et al. (2020). Correlating photons using the collective nonlinear response of atoms weakly coupled to an optical mode. *Nature Photonics*.
36. Reinhard, A., et al. (2012). Strongly correlated photons on a chip. *Nature Photonics, 6*, 93–96.
37. Reinhard, P., et al. (2021). Nuclear charge densities in spherical and deformed nuclei: Toward precise calculations of charge radii. *Physical Review C*.
38. Shalm, L. K., et al. (2012). Three-photon energy–time entanglement. *Nature Physics*.
39. Stehle, C., Zimmermann, C., & Slama, S. (2014). Cooperative coupling of ultracold atoms and surface plasmons. *Nature Physics*.
40. Szafron, R., & Czarnecki, A. (2016). High-energy electrons from the muon decay in orbit: Radiative corrections. *Physics Letters B, 753*, 61–64.
41. Tame, M. S., et al. (2013). Quantum plasmonics. *Nature Physics, 9*, 329–340.
42. Waseem, S. B. (2011). Orbital excitation blockade and algorithmic cooling in quantum gases. *Nature, 480*(7378), 500–503.

Part III
Sustainable Energy

Chapter 6
The Availability of Global Solar Energy

Introduction

There is currently 36,630 EJ of total fossil fuel reserve available on the earth, but it will soon be exhausted due to the massive energy consumption needed to power modern civilization on a global scale [12, 23]. From 283 EJ in 1980 to 348 EJ in 1995, 405 EJ in 2005, and 515 EJ in 2015, the annual consumption of fossil fuels has increased rapidly, and it is projected to reach 610 EJ in 2025, 705 EJ in 2035, 860 EJ in 2045, and 990 EJ in 2050 [11, 33]. Regardless, the conventional energy sources are still seeing fast growth in demand across the world, and this trend will likely continue until a sustainable renewable energy source is widely available [6, 9, 15]. Consequently, the current research investigates the constraints of fossil fuel reserves, with a specific focus on the fact that these sources are limited, and thus, it also investigates solar radiation as a potential alternative energy option that could be used to combat these limits in satisfying the global energy demand.

Materials and Methods

Estimation of Total Solar Energy Reaching on Earth

By defining different directional angles, the context of the Cartesian coordinate (CC) system is being introduced to better clarify the net solar irradiance reaching on overall surface area of the earth throughout the year. In this Cartesian coordinate system, the height (h) and azimuth angle (A) determine the location of the celestial body, with the central system (z hub) and its x-axis pivot to the North Pole (NP) once its y-axis is in perpendicular to the sky horizon (HZ), and its x-y axis is perpendicular to the NP and HZ [5, 7]. To apply the accurate angles to trap the solar irradiation much efficiently, the mathematical aspects of the angle and CC frequencies have

© The Author(s), under exclusive license to Springer Nature Switzerland AG 2025

M. F. Hossain, *Sustainable Global Development in Environment, Energy, Building, and Infrastructure*, https://doi.org/10.1007/978-3-031-84429-4_6

been calculated by implementing the latitude and longitude precisely [6, 9]. Here, the main meridian is the zero-degree latitude line that controls the meridian's function in the Eastern Hemisphere (ES) and the Western Hemisphere's (WS) angle of the earth's plane. Subsequently, it controls the hemispheres north of the equator in the Northern Hemisphere and also controls the hemispheres south of the equator in the Southern Hemisphere to more efficiently trap solar energy. Thus, the analytical Cartesian coordinate is being used to precisely explain, which are then used to decide where the solar irradiance vector is, which, in turn, clarifies the net amount of solar irradiance penetration on the earth [18, 21, 40]. The surface of the entire planet has been modeled considering the sun's peak radiation hours respecting the total earth's horizon considering the total areal dimension for solar irradiance reaching the Earth from the sun [19, 28]. To calculate the amount of radiation from the sun that reaches the earth is being calculated by defying the density of solar radiation flux reaches the earth per m^2/h by measuring all directional rays including perpendicular directional rays [4, 5, 7]. Then this net global solar radiation reaching the earth is being accumulated by computing all kinds of irradiance using MATLAB software.

By taking into account, the prime variable of the basic solar irradiance is being calculated and subsequently transformed it into the secondary order of variables of the movement of the photon flux of the sunlight [20, 25, 31]. Hence, the solar quantum dynamics, the most widely accepted quantum method for determining earth's net solar energy emission, is used to compute the integration of these two functions to determine the accurate solar radiation emits globally from the photon flux (W/m^2):

$$I_{\text{tot}} = I_{\text{beam}} \cos\theta + I_{\text{diffuse}}$$

Here, θ denotes the zenith-angle that is analyzed to estimate net solar irradiance penetration on the earth [3, 6, 9].

Subsequently, MATLAB Multifunctional Mapping is being incorporated to estimate the net of solar radiation penetration on earth's surface, respecting the global quantum flow of solar irradiance [3, 4, 29]. Subsequently, the correlation between the I-V-R and the earth plane is being calculated to certain the net solar energy penetration on the earth's surface [4, 8, 10]. Hence, the energy deliberate (η_{pvg}) on the earth is being calculated based on the photon irradiance and expressed as

$$\eta_{pvg} = \eta_r \eta_{pc}[1 - \beta(T_c - T_{\text{cref}})] \tag{6.1}$$

Here, η_{pc} is representing the energy formation efficiency, β stands for the variable factor, η_r represents the rate of energy penetration rate, and T_{cref} represents the temperature in °C. The solar radiation reaching on earth, which is related to the current (I) and volt relationship of the earth surface, can be written as

$$I = N_p \left[I_{ph} - I_{rs} \left[\exp\left(\frac{q(V + IR_s)}{AKTN_s} - 1 \right) \right] \right] \tag{6.2}$$

where

$$I_{rs} = I_{rr} \left(\frac{T}{T_r} \right)^3 \exp\left[\frac{E_G}{AK} \left(\frac{1}{T_r} - \frac{1}{T} \right) \right] \tag{6.3}$$

Hence, in Eqs. 6.2 and 6.3, q equals 1.6×10^{-19} C, K represents Boltzmann's constant, A represents the standard-diode efficiency, and T is the earth temperature in Kelvin. Here, I_{rs} is the reversal current at temperature T, I_{rr} is the reversal current at earth's condition temperature T_r, and E_G is the solar irradiance attenuation rate on the earth's surface. So, the solar energy-based net current I_{ph} is going to be produced based on the radiation situation that can be described as

$$I_{ph} = \left[I_{SCR} + k_i (T - T_r) \frac{S}{100} \right] \tag{6.4}$$

In this context, I_{SCR} stands for the current motion that considers the ideal earth temperature and the dynamic of solar radiation on the plane of the planet, k_i is the current motion that is short-circuited, and S is the solar power estimation in mW/m^2. So, the total solar energy (current flow) on the earth plane can be computed as

$$I = I_{PV} - I_{D1} - \left(\frac{V + IR_S}{R_{SH}} \right) \tag{6.5}$$

Calculation of Total Electricity Energy from Total Solar Radiation on the Earth

The integration of planet Albanian symmetries of gauge field is being calculated to build a model for the transformation of solar irradiance into electrical power [37, 44]. Then, the global symmetrical array of Earth's plane is being analyzed respecting the net solar irradiance dynamic to confirm the net electricity energy generation from the solar irradiance, where q represents the electric charge couples with its electromagnetic field, and the resulting equation is $\mathfrak{h}$:

$$\mathfrak{h} = -\frac{1}{4} F_{\mu\nu} F^{\mu\nu} + D_\mu \Phi^* D^\mu \Phi - V(\Phi^* \Phi) \tag{6.6}$$

where

$$D_\mu\Phi(x) = \partial_\mu\Phi(\mathrm{x}) + \mathrm{iqA_\mu(x)\Phi\ (x)}$$
$$D_\mu\Phi^*(x) = \partial_\mu\Phi^*(\mathrm{x}) - \mathrm{iqA_\mu(x)\Phi^*(x)} \tag{6.7}$$

and

$$V(\Phi^*\Phi) = \frac{\lambda}{2}\ (\Phi^*\Phi)^2 + m^2(\Phi^*\Phi) \tag{6.8}$$

Here, $\lambda > 0\ m^2 < 0$, and $\Phi = 0$ represents the local classical scalar once the maximum form of generated vector is defined as $\Phi = \frac{v}{\sqrt{2}} * e^{i\theta}$,

$$v = \sqrt{\frac{-2m^2}{\lambda}}, \text{any real } \theta \tag{6.9}$$

Then, the U (1) asymmetrical total solar power generation is being determined at the same time that the scalar field of the earth plane forms a vector forms; as a result, the following equation is being used to define the earth's U (1) total asymmetrical electricity energy of (x) as an expectant value, which is determined by the x-functional factor state of earth's surface:

$$\Phi\ (\mathrm{x}) = \frac{1}{\sqrt{2}}\ \Phi_\mathrm{r}(\mathrm{x}) * \mathrm{e}^{\mathrm{i}\Theta(\mathrm{x})}, \text{real } \Phi_\mathrm{r}(\mathrm{x}) > 0, \text{real } \Phi(\mathrm{x}) \tag{6.10}$$

As the maximum amount of solar energy is emitted from earth's surface, the net amount of electricity that can be generated from this source is calculated with the vector (x) = 0. This function is of the first order. This means that the following equation is the results of the worldwide solar energy calculation and (x) its vector on the earth plane:

$$V(\phi) = \frac{\lambda}{8}\ \left(\phi_r^2 - v^2\right)^2 + \text{const}, \tag{6.11}$$

Subsequently, total energy can be transformed by its VEV, $\Phi_r(x) = v + \sigma(x)$,

$$\phi_r^2 - v^2 = (v + \sigma)^2 - v^2 = 2v\sigma + \sigma^2 \tag{6.12}$$

$$V = \frac{\lambda}{8}\ \left(2v\sigma - \sigma^2\right)^2 = \frac{\lambda v^2}{2} * \sigma^2 + \frac{\lambda v}{2} * \sigma^3 + \frac{\lambda}{8} * \sigma^4 \tag{6.13}$$

Simultaneously, the functional variable $D_\mu\phi$ will become

$$D_\mu\phi = \frac{1}{\sqrt{2}}\ \left(\partial_\mu\left(\phi_r e^{i\Theta}\right) + iqA_\mu * \phi_r e^{i\Theta}\ \right) = \frac{e^{i\Theta}}{\sqrt{2}}\ \left(\partial_\mu\phi_r + \phi_r * i\partial_\mu\Theta + \phi_r * iqA_\mu\right) \tag{6.26}$$

$$\left|D_\mu \phi\right|^2 = \frac{1}{2}\left|\partial_\mu \phi_r + \phi_r * i\partial_\mu \Theta + \phi_r * iqA_\mu\right|^2 = \frac{1}{2}\left(\partial_\mu \phi_r\right) + \frac{\phi_r^2}{2} * \left(\partial_\mu \Theta q A_\mu\right)^2$$
$$= \frac{1}{2}\left(\partial_\mu \sigma\right)^2 + \frac{(v+\sigma)^2}{2} * \left(\partial_\mu \Theta + qA_\mu\right)^2 \tag{6.14}$$

Altogether,

$$\mathfrak{h} = \frac{1}{2}\left(\partial_\mu \sigma\right)^2 - v(\sigma) - \frac{1}{4} F_{\mu\upsilon} F^{\mu\upsilon} + \frac{(v+\sigma)^2}{2} * \left(\partial_\mu \Theta + qA_\mu\right)^2 \tag{6.15}$$

Thus, the following equation describes the net solar energy quantification on the earth plane that is used to determine the net availability of the electricity power on the earth plane due to the solar irradiance reaching the earth:

$$\mathfrak{h}_{\text{sef}} = \frac{1}{2}\left(\partial_\mu \sigma\right)^2 - \frac{\lambda v^2}{2} * \sigma^2 - \frac{1}{4} F_{\mu\upsilon} F^{\mu\upsilon} + \frac{v^2}{2} * \left(qA_\mu + \partial_\mu \Theta\right)^2 \tag{6.16}$$

Results and Discussion

Estimation of Total Solar Energy Reaching the Earth

The constant irradiance approximation technique is used by calculating the photonic irradiance array and unit area to precisely estimate the total amount of solar power reaching the earth surface [1, 16, 39]. According to the calculated results, the penetration of solar radiation on the earth plane is being estimated by the law of spinning sphere cycles in accordance with the spherical law of cosines:

$$\cos(c) = \cos(a)\cos(b) + \sin(a)\sin(b)\cos(C) \tag{6.17}$$

Here, the radius of the earth is denoted by a, b, and c, respectively. The angle of the vertex with an arc length of c is being implemented by the sun zenith angle Θ, where the spherical law of cosines clarifies the following equations:

$$C = h$$

$$c = \Theta$$

$$a = \frac{1}{2}\pi - \phi$$

$$b = \frac{1}{2}\pi - \delta$$

$$\cos(\Theta) = \sin(\phi)\sin(\delta) + \cos(\phi)\cos(\delta)\cos(h) \tag{6.18}$$

To quantify this equation more accurately, it has been further calculated as a functional derivative as follows:

$$\begin{aligned}\cos(\theta) = {} & \sin(\phi)\sin(\delta)\cos(\beta) + \sin(\delta)\cos(\phi)\sin(\beta)\cos(\gamma) \\ & + \cos(\phi)\cos(\delta)\cos(\beta)\cos(h) - \cos(\delta)\sin(\phi)\sin(\beta)\cos(\gamma)\cos(h) \\ & - \cos(\delta)\sin(\beta)\sin(\gamma)\sin(h)\end{aligned} \tag{6.19}$$

Here, β represents the angle from the horizon of the earth and γ represents the azimuth angle of the earth. As θ is representing the earth's polar angle, $\theta = 0$ at the vernal equinox and the declination δ as a function of Earth surface position can be calculated as

$$\delta = \varepsilon \sin(\theta) \tag{6.20}$$

To get a consensus of δ, θ, and ε, the Astro dynamical rules are being computed considering $Q = 0$ is thought of as the length of the earth equinox for a whole day's irradiance by implementing the following equation:

$$Q = S_0\left(1 + 0.034\cos\left(2\pi\frac{n}{365.25}\right)\right) \tag{6.21}$$

Here, n is the number of days; the solar radiation can be determined per unit area [23, 29].

To prevent the splitting from the Earth's surface, a peak high-frequency cut-off of solar radiation has been determined to conduct net calculation from the photonic energy frequencies reaching on the Earth's surface by determining the density of state (DOS) of the photons on the plane of the earth [2, 22, 24]. The depicted DOS for a one-dimensional (1D) earth plane is, thus, estimated by the Heaviside functional step [13, 17, 26]. To obtain a correct net qualitative state of solar irradiance, this DOS is, thus, committed to verifying a three-dimensional (3D) isentropic function on the earth plane to precisely calculate the net electrical energy generation as follows [14, 20, 25]:

$$J(\omega) = \varrho(\omega)|V(\omega)|^2 \tag{6.22}$$

Here, the frequency ω is repressing the photon energy mode in the photonic band gap (PBG) ($0 < \omega_b < \omega_e$), and thus, it is estimated using functional condition of $\omega_b - \omega_c - \Delta(\omega_b) = 0$, where $\lesssim\Delta(\omega) = \mathcal{P}\left[\int d\omega' \frac{J(\omega')}{\omega - \omega'}\right]$.

As a result, the net photon energy has been computed for the 1D, 2D, and 3D earth plane with regard to the PBG function, taking the proliferation magnitude into account [27, 30, 45]. The function is being used to compute the dynamic rate of solar energy. After moving from the PBG to the PB region, the results showed that a large number of photons are being released. This corresponds to crossover area that is indicated as the functional array because of the range in relation to the condition when it comes to the crossover area of the PBE of the earth plane. Hence, the generation of solar irradiance is nearly getting steady state condition; thus, this steady state photon generation confirmed that solar irradiance reaching the entire Earth plane is in a standard equilibrium condition, considering thermal fluctuation and that can be expressed as

$$(t,t) = \int_{t_0}^{t} dt_1 \int_{t_0}^{t} dt_2 u^*(t_1,t_0)\tilde{g}(t_1,t_2)u(t_2,t_0) \tag{6.25}$$

This is where the earth plane correlates the variations in solar irradiance generation due to the earth plane condition; thus, the surface's photon energy emission at its optimal temperature T can be expressed as

$$v(t, t \rightarrow \infty) = \int_{\omega_e}^{\infty} d\omega \mathcal{V}(\omega)$$

with

$$\mathcal{V}(\omega) = \overline{n}(\omega, T)[\mathcal{D}_l(\omega) + \mathcal{D}_d(\omega)] \tag{6.26}$$

By solving this aforementioned equation, it can be obtained the Fock state photon number from the photon energy, which can be obtained mathematically by considering the first photon burst (PB) on the earth at time t.

$$\rho(t) = \sum_{n=0}^{\infty} \mathcal{P}_n^{(n_0)}(t)|n_0\rangle\langle n_0| \tag{6.27}$$

$$\mathcal{P}_n^{(n_0)}(t) = \frac{[v(t,t)]^n}{[1+v(t,t)]^{n+1}}[1-\Omega(\mathrm{t})]^{n_0} \times \sum_{k=0}^{min\{n_0,n\}} \binom{n_0}{k}\binom{n}{k}\left[\frac{1}{v(t,t)}\frac{\Omega(\mathrm{t})}{1-\Omega(\mathrm{t})}\right]^k \tag{6.28}$$

where $\Omega(\mathrm{t}) = \frac{|u(t,t_0)|^2}{1+v(t,t)}$. Consequently, the outcome demonstrated that photon energy in an electron state will undergo evolution into various Fock states on earth's surface. As a result, the net solar irradiance reaching the Earth plane can be expressed as

$$\mathcal{P}_n^{(n_0)}(t \to \infty) = \frac{[\overline{n}(\omega_c, T)]^n}{[1 + \overline{n}(\omega_c, T)]^{n+1}} \tag{6.29}$$

To investigate this massive solar irradiance penetration on the earth plane, Eq. 6.29 has been implemented with photon energy as an input and computed the distribution of photon energy inside the quantum field of this surface using high-temperature coherent states. The result was expressed as

$$\rho(t) = \mathcal{D}[\alpha(t)]\rho_T[v(t,t)]\mathcal{D}^{-1}[\alpha(t)] \tag{6.30}$$

where $\mathcal{D}[\alpha(t)] = \exp\{\alpha(t)\alpha^\dagger - \alpha^*(t)\alpha\}$ denotes the displacement functions with $\alpha(t) = u(t, t_0)\alpha_0$ and

$$\rho_T[v(t,t)] = \sum_{n=0}^{\infty} \frac{[v(t,t)^n]}{[1 + v(t,t)]^{n+1}} \mid n\rangle\langle n| \tag{6.31}$$

Here, ρ_T represents the energy state photons with a mean quantum of $v(t, t)$, which implies that the net photon energy generation can be transformed from the net energy state photon by expressing the following equation:

$$\langle m|\rho(t)|n\rangle = J(\omega) = e^{-\Omega(t)|\alpha_0|^2} \frac{[\alpha(t)]^m[\alpha^*(t)]^n}{[1+v(t,t)]^{m+n+1}} = \sum_{k=0}^{\min\{m,n\}} \frac{\sqrt{m!n!}}{(m-k)!(n-k)!k!}\left[\frac{v(t,t)}{\Omega(t)|\alpha_0|^2}\right]^k \tag{6.32}$$

Calculation of Total Electricity Energy from Total Solar Radiation on the Earth

The total solar energy is then calculated using a theoretical mode of the earth plane as the circuit considering both series and parallel, to convert this immense amount of photon into electrical energy. Next, the hypothetical earth surface is integrated into the I–V single diode circuit to determine the exact I–V relationship of the net solar energy reaching the earth plane. This is done by using the following equation:

$$\mathrm{I} = \mathrm{IL} - \mathrm{IO}\left\{\exp^{\left[\frac{q(V+I_{RS})}{AkTc}\right]} - 1\right\} - \frac{(V + I_{RS})}{R_{Sh}} \tag{6.33}$$

In this context, I_L stands for the current that forms photons, represents the ideal current flow, Rs represents the series resistance, A represents the diode function, k is Boltzmann's constant, q represents the electron charge, and TC represents the temperature of the earth [32, 34, 47].

Hence, there are a number of different unit areas and self-energy-induced reservoirs that correlate to the many modes of solar radiation that reach the earth's surface [35, 43, 46]. Taking into account this hypothetical surface of the earth, the I–V relationship has been calculated for all of the hypothetical solar cells by modifying Eq. 6.34 as follows:

$$V = -IR_s + K \log\left[\frac{I_L - I + I_O}{I_O}\right] \tag{6.34}$$

In this case, K stands for the constant, while I_{mo} and V_{mo} represent the net current and voltage on the hypothetical surface of the earth, respectively. Consequently, on the I–V earth surface, the interaction between I_{mo} and V_{mo} will continue to be dynamic, and this can be expressed as

$$V_{mo} = -I_{mo}R_{Smo} + K_{mo} \log\left(\frac{I_{Lmo} - I_{mo} + I_{Omo}}{I_{Omo}}\right) \tag{6.35}$$

As the current flow into the circuit is lined up with the cells, the series resistance is calculated by adding up the series resistance of all the non-series (NS) cells, which is $R_{smo} = NS \times RS$ current, taking into account the mode of the consistent factor $K_{mo} = NS \times K$. This process is repeated for each solar cell [36, 38, 41]. Therefore, the mode of the I_{mo}–V_{mo} relationship for the NS series of connected cells can be described by and the current dynamics as

$$V_{mo} = -I_{mo}N_SR_S + N_SK\log\left(\frac{I_L - I_{mo} + I_o}{I_o}\right) \tag{6.36}$$

Subsequently, the I–V relationship can be further addressed considering all parallel linked into N_P cells connection in all parallel mode as follows [42, 48]:

$$V_{mo} = -I_{mo}\frac{Rs}{Np} + K\log\left(\frac{N_{sh}I_L - I_{mo} + N_pI_o}{N_pI_o}\right) \tag{6.37}$$

Finally, the net emitted photon utilization can be verified by computing the net electricity power formation emission over the Earth's surface and integrating local Albanian electric fields of the transformation law of a scalar field as [22, 44]:

$$\begin{aligned} \partial_\mu \rightarrow D_\mu = \partial_\mu = ieA_\mu \quad & [\text{covariant derivatives}] \\ A'_\mu = A_\mu + \frac{1}{e}\,\partial_\mu\alpha \quad & [A_\mu \text{ derivatives}] \end{aligned} \tag{6.39}$$

Here, the global U (1) gauge denotes the invariant local Albanian for a complex scalar field which further can be expressed as

$$\mathfrak{h} = (D^{\mu})^{\dagger}\left(D_{\mu}\varnothing\right) - \frac{1}{4}F_{\mu\nu}F^{\mu\nu} - V(\varnothing) \tag{6.40}$$

The term represents the additional term in the local Albanian language and is the dynamic name for the earth's surface gauge field.

Thus, the computation of mass-scalar particles and a mass variable with the perturbational function of the earth plane has validated by the production of local Albanian ($\mathfrak{h}$). In this state, there were an endless amount of quantum fluctuations, which are being explained by the variable derivatives and the subsequent shifting of fields, which characterized the quantum field as

$$\text{Kinetic term}: \mathfrak{h}(\eta,\xi) = (D^{\mu}\phi)^{\dagger}(D^{\mu}\phi) = (\partial^{\mu} + ieA^{\mu})\phi^{*}\left(\partial_{\mu} - ieA_{\mu}\right)\phi \tag{6.41}$$

Therefore, the net amount electricity energy generation in the $\mathfrak{h}$ is related to the earth's scalar field which is the entire plane or the earth's plane surface [4, 8, 11].

This electrical energy generation is the final non-variable function of energy dynamics, which can be expressed as follows:

$$\overline{\phi}[s_0] = 2s_0(\ln s_0 - 2) + \ln 4s_0(\ln 4s_0 - 2)\left(\frac{\pi^2 - 9}{3}\right) + s_0^{-1}\left(\ln 4s_0 + \frac{9}{8}\right) + \ldots(s_0 \gg 1); \tag{6.42}$$

$$\overline{\phi}[s_0] = \left(\frac{2}{3}\right)(S_0 - 1)^{\frac{3}{2}} + \left(\frac{5}{3}\right)(S_0 - 1)^{\frac{5}{2}} - \left(\frac{1507}{420}\right)(S_0 - 1)^{\frac{7}{2}}(1/2 \text{ instead of } 1). \tag{6.43}$$

Then the final equation can be rewritten as where s_0 is the areal value of electricity energy generation into the Earth surface (1 m^2).

$$\overline{\phi}[s_0] = \left(\frac{2}{3}\right)(S_0 - 1)^{\frac{3}{2}} + \left(\frac{5}{3}\right)(S_0 - 1)^{\frac{5}{2}} - \left(\frac{1507}{420}\right)(S_0 - 1)^{\frac{7}{2}} \tag{6.44}$$

Based on the estimate of earth's cross-sectional area, which is 127,400,000 km^2, the net solar energy intercepted by the plane of the planet is 1.740 × 1017 Watts, and the net electricity energy generation from the total solar energy is, thus, given by the function of [s0] that reaches the ground on a daily basis is typically 200 Watts per square meter [2, 32]. The world consumed energy in 2023 was 22,126 TWh [4, 12]. Hence, the total power that reaches on earth's surface at every second is computed as 222,504,000 (TWh) or 25,400 terawatts (TW) by multiplying TW by 24 by 365 which is more than 10,000 times the net annual energy worldwide.

Conclusion

The modern world is facing a serious challenge in the form of the increasing consumption of fossil fuels, which is not only contributing to the depletion of climate change but is also approaching a finite level in the near future. Thus, meeting the energy demands of the world requires the adoption of an abundant and dependable alternative energy source. Interestingly, the most potential solution to the energy demand for the world is indeed solar power since it is clean, renewable, and abundant everywhere. This study suggested that the net energy demand of the entire world can be met with a fraction of 0.001% solar energy that enters the Earth's surface each year which is clean and environmentally friendly. By harnessing its abundant and eco-friendly nature, this study aims to secure energy security for all individuals in the near future and tackle the consequences of climate change and energy shortage crisis in the future.

Acknowledgments The author confirms that he is solely responsible for all predictions, forecasts, and results stated in this chapter and the chapter bears no conflict of interest for publishing any standard publisher.

References

1. Armani, D. K., Kippenberg, T. J., Spillane, S. M., & Vahala, K. J. (2003). Ultra-high-Q toroid microcavity on a chip. *Nature, 421*(6926), 925–928.
2. Baur, G., Hencken, K., & Trautmann, D. (2007). Revisiting unitarity corrections for electromagnetic processes in collisions of relativistic nuclei. *Physics Reports, 453*(1).
3. Ben Messaoud, R. (2019). Extraction of uncertain parameters of double-diode model of a photovoltaic panel using simulated annealing optimization. *The Journal of Physical Chemistry C, 123*(48), 29096–29103.
4. Besharat, F., Dehghan, A. A., & Faghih, A. R. (2013). Empirical models for estimating global solar radiation: A review and case study. *Renewable and Sustainable Energy Reviews, 21*, 798–821.
5. Birnbaum, K. M., Boca, A., Miller, R., Boozer, A. D., Northup, T. E., & Kimble, H. J. (2005). Photon blockade in an optical cavity with one trapped atom. *Nature, 436*(7047), 87–90.
6. Brešar, B., Klavžar, S., & Škrekovski, R. (2003). Quasi-median graphs, their generalizations, and tree-like equalities. *European Journal of Combinatorics, 24*(5), 557–572.
7. Chang, D. E., Sørensen, A. S., Demler, E. A., & Lukin, M. D. (2007). A single-photon transistor using nanoscale surface plasmons. *Nature Physics, 3*(11), 807–812.
8. Danish, M. S. S., & Senjyu, T. S. (2020). Green building efficiency and sustainability indicators. In *Green building management and smart automation* (pp. 128–145). IGI Global.
9. Dayan, B., Parkins, A. S., Aoki, T., Ostby, E. P., Vahala, K. J., & Kimble, H. J. (2008). A photon turnstile dynamically regulated by one atom. *Science, 319*(5866), 1062–1065.
10. Douglas, J. S., Habibian, H., Hung, C., Gorshkov, A. V., Kimble, H. J., & Chang, D. E. (2015). Quantum many-body models with cold atoms coupled to photonic crystals. *Nature Photonics, 9* (5), 326–331.
11. Evaristo, J., Jasechko, S., & McDonnell, J. J. (2015). Global separation of plant transpiration from groundwater and streamflow. *Nature, 525*(7567), 91–94.

12. Fratzl, P., & Barth, F. G. (2009). Biomaterial systems for mechanosensing and actuation. *Nature, 462*(7272), 442–448.
13. Gleyzes, S., Kuhr, S., Guerlin, C., Bernu, J., Deleglise, S., Busk Hoff, U., et al. (2007). Quantum jumps of light recording the birth and death of a photon in a cavity. *Nature, 446* (7133), 297–300.
14. Guerlin, C., Bernu, J., Deleglise, S., Sayrin, C., Gleyzes, S., Kuhr, S., et al. (2007). Progressive field-state collapse and quantum non-demolition photon counting. *Nature, 448*(7156), 889–893.
15. Guo, Y., Al-Jubainawi, A., & Ma, Z. (2019). Performance investigation and optimisation of electrodialysis regeneration for LiCl liquid desiccant cooling systems. *Applied Thermal Engineering, 149*, 1023–1034.
16. Hossain, M. F. (2016). Solar energy integration into advanced building design for meeting energy demand and environment problem. *International Journal of Energy Research, 40*(9), 1293–1300.
17. Hossain, M. F. (2017). Green science: Independent building technology to mitigate energy, environment, and climate change. *Renewable and Sustainable Energy Reviews, 73*, 695–705.
18. Hossain, M. (2018). Transforming dark photons into sustainable energy. *International Journal of Energy and Environmental Engineering, 9*(1), 99–110.
19. Hossain, M. F. (2018a). Green science: Advanced building design technology to mitigate energy and environment. *Renewable and Sustainable Energy Reviews, 81*, 3051–3060.
20. Hossain, M. F. (2018b). Photon energy amplification for the design of a micro PV panel. *International Journal of Energy Research, 42*(12), 3861–3876.
21. Hossain, M. (2019). Green technology: Transformation of transpiration vapor to mitigate global water crisis. *Polytechnica, 2*(1), 26–29.
22. Hossain, M. F. (2019a). Advanced building design. *Sustainable Design and Build.*
23. Hossain, M. F. (2019b). *Applied energy technology.*
24. Hossain, M. F. (2019c). No title. RETRACTED: *Natural mechanism to console global water, energy, and climate change crisis.*
25. Hossain, M. F. (2019d). No title. RETRACTED: *Theoretical mechanism to breakdown of photonic structure to design a micro PV panel.*
26. Hossain, M. F. (2019e). Sustainable technology for energy and environmental benign building design. *Journal of Building Engineering, 22*, 130–139.
27. Hossain, M. F. (2020). No title. *Photonic thermal control to naturally cool and heat the building* (Retraction of Vol 131, Pg 576, 2018).
28. Hossain, M. (2022). Photon application in the design of sustainable buildings to console global energy and environment. In *Sustainable design for global equilibrium* (pp. 125–142). Springer.
29. Hossain, M. F., & Pota, H. R. (2019). *Power systems.*
30. Jasechko, S., Sharp, Z. D., Gibson, J. J., Birks, S. J., Yi, Y., & Fawcett, P. J. (2013). Terrestrial water fluxes dominated by transpiration. *Nature, 496*(7445), 347–350.
31. Joannopoulos, J. D., Villeneuve, P. R., & Fan, S. (1997). Photonic crystals: Putting a new twist on light. *Nature, 386*(6621), 143–149.
32. Langer, L., Poltavtsev, S. V., Yugova, I. A., Salewski, M., Yakovlev, D. R., Karczewski, G., et al. (2014). Access to long-term optical memories using photon echoes retrieved from semiconductor spins. *Nature Photonics, 8*(11), 851–857.
33. Maxwell, R. M., & Condon, L. E. (2016). Connections between groundwater flow and transpiration partitioning. *Science, 353*(6297), 377–380.
34. Newell, J. R. (2009). A story of things yet-to-be: The status of geology in the United States in 1807. *Geological Society, London, Special Publications, 317*(1), 213–217.
35. Park, K., Marek, P., & Filip, R. (2017). Qubit-mediated deterministic nonlinear gates for quantum oscillators. *Scientific Reports, 7*(1), 1–8.
36. Reinhard, A., Volz, T., Winger, M., Badolato, A., Hennessy, K. J., Hu, E. L., & Imamoğlu, A. (2012). Strongly correlated photons on a chip. *Nature Photonics, 6*(2), 93–96.

37. Sayrin, C., Dotsenko, I., Zhou, X., Peaudecerf, B., Rybarczyk, T., Gleyzes, S., & Brune, M. (2011). Real-time quantum feedback prepares and stabilizes photon number states. *Nature, 477*(7362), 73–77.
38. Shen, H. Z., Xu, S., Cui, H. T., & Yi, X. X. (2019). Non-markovian dynamics of a system of two-level atoms coupled to a structured environment. *Physical Review A, 99*(3), 032101.
39. Tame, M. S., McEnery, K. R., Özdemir, Ş. K., Lee, J., Maier, S. A., & Kim, M. S. (2013). Quantum plasmonics. *Nature Physics, 9*(6), 329–340.
40. Tu, M. W., & Zhang, W. (2008). Non-markovian decoherence theory for a double-dot charge qubit. *Physical Review B, 78*(23), 235311.
41. Vargas, R., Carvajal, D., Madriz, L., & Scharifker, B. R. (2020). Chemical kinetics in solar to chemical energy conversion: The photoelectrochemical oxygen transfer reaction. *Energy Reports, 6*, 2–12.
42. Wheeler, T. D., & Stroock, A. D. (2008). The transpiration of water at negative pressures in a synthetic tree. *Nature, 455*(7210), 208–212.
43. Xiao, Y., Li, M., Liu, Y., Li, Y., Sun, X., & Gong, Q. (2010). Asymmetric fano resonance analysis in indirectly coupled microresonators. *Physical Review A, 82*(6), 065804.
44. Yan, W., & Fan, H. (2014). Single-photon quantum router with multiple output ports. *Scientific Reports, 4*(1), 1–6.
45. Yan, W., Huang, J., & Fan, H. (2013). Tunable single-photon frequency conversion in a sagnac interferometer. *Scientific Reports, 3*(1), 1–6.
46. Yang, L., Wang, S., Zeng, Q., Zhang, Z., Pei, T., Li, Y., & Peng, L. (2011). Efficient photovoltage multiplication in carbon nanotubes. *Nature Photonics, 5*(11), 672–676.
47. Zhang, W., Lo, P., Xiong, H., Tu, M. W., & Nori, F. (2012). General non-markovian dynamics of open quantum systems. *Physical Review Letters, 109*(17), 170402.
48. Zhu, Y., Hu, X., Yang, H., & Gong, Q. (2014). On-chip plasmon-induced transparency based on plasmonic coupled nanocavities. *Scientific Reports, 4*(1), 1–7.

Chapter 7
Remodeling of Photon Structure to Produce Renewable Energy

Introduction

This study presents a novel micro photovoltaic cell to improving the efficiency solar irradiance trapping by converting one photon into several photons to address the global power crisis completely. Hence, the nanoscale photon-band-gap (PBG) structures of the photons are being clarified using MATLAB optical-circuit former, in conjunction with it, nanoscale point breaks to agitate the dormant photon into an energetically active state photon to prove that point defects can generate much photon from a single photon [3, 7, 9]. Subsequently, this approach is being to combine photovoltaic cell with photon-band-gap (PBG) waveguide that defect the arrays of the cells to detect photon quantum dynamics by measuring point defect waveguides from the derivative of the dormant B–E photon theory [12, 36, 37]. Naturally, under these circumstances, the inactive photon state is being analyzed to deconstruct the photonic structure by using MATLAB/Simulink for the goal of developing a micro photovoltaic cell to emit much more solar energy [1, 16, 44]. Thus, a thorough mathematical calculation is computed to clarify the concept of an advanced photovoltaic cell system development that PV current source and a four-diode circuit to calculate the efficiency of this solar cell perfectly to address the net energy requirement for the world.

Methods and Simulations

The photon dynamics considering extreme relativistic state were computed using photonic waveguides embedded in an acting photovoltaic cell to consider it as the reservoirs of the active photons (HeP^{-}). Thus, for this specific computation, the contour maps of the photon have been determined by the following equation:

© The Author(s), under exclusive license to Springer Nature Switzerland AG 2025

M. F. Hossain, *Sustainable Global Development in Environment, Energy, Building, and Infrastructure*, https://doi.org/10.1007/978-3-031-84429-4_7

$$\rho(t) = -i\left[H_c'(t), \rho(t)\right] + \sum_{ij}\left\{k_{ij}(t)\left[2a_j\rho(t)a_i^\dagger - a_i^\dagger a_j\rho(t) - \rho(t)a_i^\dagger a_j\right]\right. \quad (7.2)$$
$$\left. + k_{ij}(t)\left[a_i^\dagger\rho(t)a_j + a_j\rho(t)a_i^\dagger - a_i^\dagger a_j\rho(t) - \rho(t)a_ja_i^\dagger\right]\right\}$$

Here, is the density of photons attenuated at the breakpoint is the re-formatted frequencies of the photons [2, 5, 10]. Subsequently, a photonic dynamic inside the photovoltaic semiconductor is being clarified considering photon quantum dynamics using the following formulas:

$$\omega_c'(t) = -Im\left[u\,(t, t_0)/u\,(t, t_0)\right] \quad (7.3)$$

Here, the photonic region of the point break represents the photon dynamics and thus, the theory of nonequilibrium dynamics can be explained as the function of

$$J\,(\omega) = \sum_k |V_k|^2\,\delta\,(\omega - \omega_k) = \varrho\,(\omega)\mid V\,(\omega)|^2 = [n * e(1 + 2n)]^4 \quad (7.8)$$

Finally, the breakpoint is being clarified by summarizing the expansion of photon formation production considering the photon dynamic state inside the photovoltaic cell by employing a four-diode cell [13, 36, 37]. Therefore, the photon formation is being precisely determined by the creation of the nonequilibrium photon as

$$J\,(\omega) = [n * e(1 + 2n)]^4 \quad (7.9)$$

Then, a point break model PV panel is being used to convert into I–V relationship that is written as

$$\mathrm{I} = \mathrm{IL} - \mathrm{I}_O\left\{\exp^{\left[\frac{q(V+I_{RS})}{AkTc}\right]} - 1\right\} - \frac{(V + I_{RS})}{R_{Sh}} \quad (7.10)$$

Here, I_L stands for photon creating energy, Rs stands for resistance, A stands for the four diodes' functional array, k is equal to 1.38×10^{-23} W/m^{2}K, q stands for electron charge, and T_c stands for active cell temperature [14, 32, 33]. Subsequently, the I–q connection changes in solar cells as a function of the four-diode energy can be written as

$$\mathrm{I}_O = \mathrm{I}_{\mathrm{RS}}\left(\frac{T_C}{T_{\mathrm{ref}}}\right)^3 \exp\left[\frac{qEG\left(\frac{1}{T_{\mathrm{ref}}} - \frac{1}{T_C}\right)}{KA}\right] \quad (7.11)$$

Here, the photon-generated current (Io), saturation current (IRS) flowing into the diode, series resistance (RS), and fractional constant (KA) are all defined based on volts (V0) and expressed as solar energy in KW/m^2.

Results and Discussion

The mathematical determination of HeP^- production in the PV cell has been conducted to confirm the photon generation occurred due to point break dynamics in the photovoltaic cell considering the different dynamic photon conditional mode of 1D: One Dimensional, 2D: Two Dimensional, 3D: Three Dimensional at photonic density (DOS) are being clarified in Table 7.1.

Hence, a high-frequency fine-cut-off is used to prevent DOS bifurcation in 3D solar cells and piroot to that a positive DOS is maintained in 2D and 1D solar cells by using of a sharper cut-off of high-level frequency in consideration of the photovoltaic nanostructure to compute the DOS of several photovoltaic cells accurately [21, 23]. Then the DOS formation array in the solar cell is described as the unit area as per the following equation:

$$J(\omega) = Q(\omega)|V(\omega)|^2 \tag{7.17}$$

Hence, it has been confirmed that the formation of photon dynamics due to point break dynamics in the photovoltaic cell considering the different dynamic photon conditional mode of 1D, 2D, 3D [15, 27, 38]. Thus, the photon band frequency here ω_c is being used for the DOS as $u(t, t_0)$ for the structural analysis of photon as:

$$(t, t_0) = \frac{1}{1 - \Sigma'(\omega_b)} e^{-i\omega(t - t_0)} + \int_{\omega_e}^{\infty} d\omega \frac{J(\omega) e^{-i\omega(t - t_0)}}{[\omega - \omega_c - \Delta(\omega)]^2 + \pi^2 J^2(\omega)} \tag{7.18}$$

$\Sigma'(\omega_b) = [\partial\Sigma(\omega)/\partial\omega]_{\omega=\omega_b}$, and $\Sigma(\omega)$ is representing here the photon self-energy generation that can be further classified as

$$\Sigma(\omega) = \int_{\omega_e}^{\infty} d\omega' \frac{J(\omega')}{\omega - \omega'} \tag{7.19}$$

Hence, the function ω_b in Eq. (7.18) represents the function of photon frequency in PBG, and thus, PBG ($0 < \omega_b < \omega_e$) can be estimated by using the following equation:

Table 7.1 The photo voltaic (PV) production in different DOS dimensional modes per unit area. Hence, the photonic structures in the below table correspond to the photovoltaic of 3D, 2D, and 1D in the photovoltaic cell.

Photo voltaic (PV)	Unit area $J(\omega)$ for various photonic density (DOS)	HeP^{-} state energy production $\Sigma(\omega)$
1D	$\frac{C}{\pi}\frac{1}{\sqrt{\omega-\omega_e}}\Theta(\omega-\omega_e)$	$-\frac{C}{\sqrt{\omega_e-\omega}}$
2D	$-\eta\left[\ln\left\|\frac{\omega-\omega_0}{\omega_0}\right\|-1\right]\Theta(\omega-\omega_e)\Theta(\Omega_d-\omega)$	$\eta\left[\mathrm{Li}_2\left(\frac{\Omega_d-\omega_0}{\omega-\omega_0}\right)-\mathrm{Li}_2\left(\frac{\omega_0-\omega_e}{\omega_0-\omega}\right)-\ln\frac{\omega_0-\omega_e}{\Omega_d-\omega_0}\ln\frac{\omega_e-\omega}{\omega_0-\omega}\right]$
3D	$\chi\sqrt{\frac{\omega-\omega_e}{\Omega_C}}\exp\left(-\frac{\omega-\omega_e}{\Omega_C}\right)\Theta(\omega-\omega_e)$	$\chi\left[\pi\sqrt{\frac{\omega_e-\omega}{\Omega_C}}\exp\left(-\frac{\omega-\omega_e}{\Omega_C}\right)\mathrm{erfc}\sqrt{\frac{\omega_e-\omega}{\Omega_C}}-\sqrt{\pi}\right]$

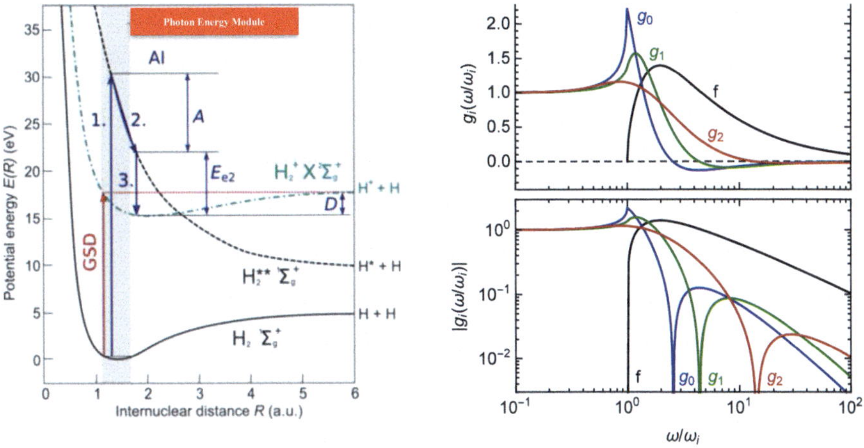

A B

Fig. 7.1 (**a**) I–V explanation of the four-diode setup with a default state to form energy by taking solar irradiance into account. (**b**) Irradiance as a photonic energy component, with the black line representing the area that absorbs photons. The dispersion zone of associated solar energy is represented by the blue line. The green line represents the average angle for the isotropic dispersion of photonic irradiance with respect to static energy

$$\omega_b - \omega_c - \Delta(\omega_b) = 0,$$

with a prime integrated function of$\lesssim \Delta(\omega) = \mathcal{P}\left[\int d\omega' \frac{J(\omega')}{\omega - \omega'}\right]$.

In addition, taking the photon formation rate into account, the photonic dynamics in 3D, 2D, and 1D have been calculated considering the photonic dynamic is being determined by incorporating different detuning from the photonic band gap area to the photonic band (PB) array [4, 6, 8]. Hence, the computed result shows that the rate of creating dynamic photons is high in the PB area. Consequently, it is possible to characterize the crossing area as linked to the condition because of the range of the photonic band gap (PBG) and the production rate in that region [29, 30, 37]. This significant variance of the dynamic of photon formation rate is due to the tiny nonequilibrium photonic band frequency in the crossover area (*see* Fig. 7.1).

Thus, the photon proliferation is being analyzed in consideration of thermal functions $v(t,t)$ and expressed:

$$v(t,t) = \int_{t_0}^{t} dt_1 \int_{t_0}^{t} dt_2 u^*(t_1,t_0)\tilde{g}(t_1,t_2)u(t_2,t_0) \qquad (7.20)$$

Hence, by reformatting Eq. 7.20 it can be regenerated as $\tilde{g}(t_1, t_2) = \int d\omega J(\omega)\sim \tilde{n}(\omega, T)e^{-i\omega(t-t')}$ to confirm the functional array of the photonic dynamic as $\overline{n}(\omega, T) = 1/\left[e^{\hbar\omega/k_B T} - 1\right]$ and it is expressed as follows:

$$v(t, t \to \infty) = \int_{\omega_e}^{\infty} d\omega \mathcal{V}(\omega),$$

with

$$\mathcal{V}(\omega) = \overline{n}(\omega, T)[\mathcal{D}_l(\omega) + \mathcal{D}_d(\omega)] \tag{7.21}$$

With respect to Eq. (7.21), this photonic band does certainly support the dynamical viability; thus, the nonequilibrium production of proliferating photons is confirmed by plotting the dynamic photon counts using following equations:

$$\rho(t) = \sum_{n=0}^{\infty} p_n^{(n_0)}(t)|n_0\rangle\langle n_0| \tag{7.22}$$

$$p_n^{(n_0)}(t) = \frac{[v(t, t)]^n}{[1 + v(t, t)]^{n+1}}[1 - \Omega(\mathrm{t})]^{n_0} \text{ x } \sum_{k=0}^{\min\{n_0 n\}} \binom{n_0}{k}\binom{n}{k}\left[\frac{1}{v(t, t)}\frac{\Omega(\mathrm{t})}{1 - \Omega(\mathrm{t})}\right]^k \tag{7.23}$$

The implication here is that photons in the Fock state will undergo transformation into other Fock states [19, 25, 26]. Thus, it confirms that the photon proliferation dissipation in both prime and steady regimes will experience thermal nonequilibrium which is expressed as

$$J(\omega) = e^{-\Omega(t)|\alpha_0|^2}\frac{[\alpha(t)]^m[\alpha^*(t)]^n}{[1 + v(t, t)]^{m+n+1}} = \sum_{k=0}^{\min\{m, n\}} \frac{\sqrt{m!n!}}{(m-k)!(n-k)!k!}\left[\frac{v(t, t)}{\Omega(t)|\alpha_0|^2}\right]^k \tag{7.24}$$

In this case, the photovoltaic panel's HeP^- production will definitely enter a nonequilibrium state, and thus, it indicates that the photonic band does not follow the *B–E* mechanism, but rather from photons exponentially [11, 17, 42]. Thus, it is possible to convert photons into on-equilibrium state photons with regard to DOS photon activation by utilizing extreme relativistic states and a clear explanation of the dynamic photon, which essentially breaks down the dormant *B–E* photons and transforms it into many photons inside the photovoltaic panels [35, 38, 40]. Hence, the model suggests that a four-diode photovoltaic cell, which consists of a tiny semiconductor connected to a circuit of silicon films, has a great function of converting high-energy photons (HeP^-) into electrical energy [20, 24]. Subsequently, the ideal photovoltaic panel would have maximum current and voltage, and an extreme power point can be clarified by using maximum current and instantaneous voltage calculations when nonequilibrium photons are captured [28, 45, 46]. Thus, at I_{mp} and V_{mp}, the power that will be delivered by a solar panel will reach its maximum value [18, 31, 41]. Consequently, this leads to the total photon formation from a single photon would be 81 by the employing the question 9, and thus, the estimated highest amount of solar radiation is produced by 81 photons as 1.4 eV × 81 = 113.4 eV (one photon is 1.4 eV). So, the total energy would be 2249.37 mW/m^2—eV per hour, which is equal to 27.77 mW/m^2 multiplied by 81, or 1.4 eV. Thus, the total energy production amounts would be 4,105,100.25 kW/year, according to the calculation, when averaged over 365 days at a maximum of 5 h each day [34, 39, 43]. This translates to 11,246.85 kW of daily energy output from a 1 m^2 solar panel. To meet the total energy demand for a 32 m × 31 m footprint and 30 m (10 story) high commercial building consumes about 3800 kWh/day would require only 0.00008% of the exterior curtain needed to be use as the acting PV panel to meet the building total energy demand.

Conclusions

The excellent reformation of *B–E* photon distribution mechanism to create high-energy photon (HeP^-) to produce clean energy would indeed be the most interesting discovery in the energy science. Hence, the formation of massive HeP^- particles from a single B–E quiescent photon in the PV cell has been verified by a set of mathematical calculations, with extreme computational conditions. The result suggested that at a very nonequilibrium conditions, the Bose–Einstein inactive photons break down, leading to an almost exponential rate of photon generation. Thus, this change takes place as the whole mechanism moves from a standard to un-equilibrium condition. As a result, the formation of these photons is enhanced by the extreme un-equilibrium condition, which allows for photonic interactions inside the PV cells and leads to an exponential growth in photon production that can be transformed into the photovoltaic (PV) circuit system to produced clean energy naturally. A thorough evaluation of the photovoltaic (PV) cell's ability to generate electricity is confirmed by a set of mathematical analyses which suggested that it is quite feasible to meet a building's entire energy demand with renewable power sources by converting as little as 0.00008% of its exterior curtain wall into an advanced solar photovoltaic energy cell. By utilizing this technology, a new era of scientific progress might be ushered in, with the dual goals of meeting the whole energy demand of the construction industry and significantly reducing the impact of climate change.

Acknowledgments The author declares that any findings, predictions, and conclusions described in this chapter are solely performed by him.

References

1. Armani, D. K., Kippenberg, T. J., Spillane, S. M., & Vahala, K. J. (2003). Ultra-high-Q toroid microcavity on a chip. *Nature, 421*(6926), 925–928.
2. Birnbaum, K. M., Boca, A., Miller, R., Boozer, A. D., Northup, T. E., & Kimble, H. J. (2005). Photon blockade in an optical cavity with one trapped atom. *Nature, 436*(7047), 87–90.
3. Busch, K., Von Freymann, G., Linden, S., Mingaleev, S. F., Tkeshelashvili, L., & Wegener, M. (2007). Periodic nanostructures for photonics. *Physics Reports, 444*(3–6), 101–202.
4. Chang, D. E., Sørensen, A. S., Demler, E. A., & Lukin, M. D. (2007). A single-photon transistor using nanoscale surface plasmons. *Nature Physics, 3*(11), 807–812.
5. Chen, J., Wang, C., Zhang, R., & Xiao, J. (2012). Multiple plasmon-induced transparencies in coupled-resonator systems. *Optics Letters, 37*(24), 5133–5135.
6. Cheng, M., & Song, Y. (2012). Fano resonance analysis in a pair of semiconductor quantum dots coupling to a metal nanowire. *Optics Letters, 37*(5), 978–980.
7. Dayan, B., Parkins, A. S., Aoki, T., Ostby, E. P., Vahala, K. J., & Kimble, H. J. (2008). A photon turnstile dynamically regulated by one atom. *Science, 319*(5866), 1062–1065.
8. Douglas, J. S., Habibian, H., Hung, C., Gorshkov, A. V., Kimble, H. J., & Chang, D. E. (2015). Quantum many-body models with cold atoms coupled to photonic crystals. *Nature Photonics, 9* (5), 326–331.
9. Englund, D., Majumdar, A., Faraon, A., Toishi, M., Stoltz, N., Petroff, P., & Vučković, J. (2010). Resonant excitation of a quantum dot strongly coupled to a photonic crystal nanocavity. *Physical Review Letters, 104*(7), 073904.
10. Gleyzes, S., Kuhr, S., Guerlin, C., Bernu, J., Deleglise, S., Busk Hoff, U., et al. (2007). Quantum jumps of light recording the birth and death of a photon in a cavity. *Nature, 446* (7133), 297–300.
11. Guerlin, C., Bernu, J., Deleglise, S., Sayrin, C., Gleyzes, S., Kuhr, S., et al. (2007). Progressive field-state collapse and quantum non-demolition photon counting. *Nature, 448*(7156), 889–893.
12. Han, Z., & Bozhevolnyi, S. I. (2011). Plasmon-induced transparency with detuned ultracompact fabry-perot resonators in integrated plasmonic devices. *Optics Express, 19*(4), 3251–3257.
13. Hossain, M. F. (2016). Solar energy integration into advanced building design for meeting energy demand and environment problem. *International Journal of Energy Research, 40*(9), 1293–1300.
14. Hossain, M. F. (2017a). Design and construction of ultra-relativistic collision PV panel and its application into building sector to mitigate total energy demand. *Journal of Building Engineering, 9*, 147–154.
15. Hossain, M. F. (2017b). Green science: Independent building technology to mitigate energy, environment, and climate change. *Renewable and Sustainable Energy Reviews, 73*, 695–705.
16. Joannopoulos, J. D., Villeneuve, P. R., & Fan, S. (1997). Photonic crystals: Putting a new twist on light. *Nature, 386*(6621), 143–149.
17. John, S., & Wang, J. (1991). Quantum optics of localized light in a photonic band gap. *Physical Review B, 43*(16), 12772.
18. Kofman, A. G., Kurizki, G., & Sherman, B. (1994). Spontaneous and induced atomic decay in photonic band structures. *Journal of Modern Optics, 41*(2), 353–384.
19. Kolchin, P., Oulton, R. F., & Zhang, X. (2011). Nonlinear quantum optics in a waveguide: Distinct single photons strongly interacting at the single atom level. *Physical Review Letters, 106*(11), 113601.
20. Lang, C., Bozyigit, D., Eichler, C., Steffen, L., Fink, J. M., Abdumalikov, A. A., Jr., et al. (2011). Observation of resonant photon blockade at microwave frequencies using correlation function measurements. *Physical Review Letters, 106*(24), 243601.
21. Lei, C. U., & Zhang, W. (2012). A quantum photonic dissipative transport theory. *Annals of Physics, 327*(5), 1408–1433.
22. Li, Q., Xu, D. Z., Cai, C. Y., & Sun, C. P. (2013). Recoil effects of a motional scatterer on single-photon scattering in one dimension. *Scientific Reports, 3*(1), 1–6.

23. Liao, J., & Law, C. K. (2010). Correlated two-photon transport in a one-dimensional waveguide side-coupled to a nonlinear cavity. *Physical Review A, 82*(5), 053836.
24. Liao, J., & Law, C. K. (2013). Correlated two-photon scattering in cavity optomechanics. *Physical Review A, 87*(4), 043809.
25. Lo, P., Xiong, H., & Zhang, W. (2015). Breakdown of bose-einstein distribution in photonic crystals. *Scientific Reports, 5*(1), 1–9.
26. Longo, P., Schmitteckert, P., & Busch, K. (2011). Few-photon transport in low-dimensional systems. *Physical Review A, 83*(6), 063828.
27. Lü, X., Zhang, W., Ashhab, S., Wu, Y., & Nori, F. (2013). Quantum-criticality-induced strong kerr nonlinearities in optomechanical systems. *Scientific Reports, 3*(1), 1–9.
28. Mahan, G. D. (2000). *Many-body physics*.
29. Martens, C., Longo, P., & Busch, K. (2013). Photon transport in one-dimensional systems coupled to three-level quantum impurities. *New Journal of Physics, 15*(8), 083019.
30. Noda, S., & Baba, T. (2003). *Roadmap on photonic crystals*. Springer.
31. O'Shea, D., Junge, C., Volz, J., & Rauschenbeutel, A. (2013). Fiber-optical switch controlled by a single atom. *Physical Review Letters, 111*(19), 193601.
32. Reinhard, A., Volz, T., Winger, M., Badolato, A., Hennessy, K. J., Hu, E. L., & Imamoğlu, A. (2012). Strongly correlated photons on a chip. *Nature Photonics, 6*(2), 93–96.
33. Roy, D. (2013). Two-photon scattering of a tightly focused weak light beam from a small atomic ensemble: An optical probe to detect atomic level structures. *Physical Review A, 87*(6), 063819.
34. Saloux, E., Teyssedou, A., & Sorin, M. (2011). Explicit model of photovoltaic panels to determine voltages and currents at the maximum power point. *Solar Energy, 85*(5), 713–722.
35. Shen, J., & Fan, S. (2007). Strongly correlated two-photon transport in a one-dimensional waveguide coupled to a two-level system. *Physical Review Letters, 98*(15), 153003.
36. Shi, T., Fan, S., & Sun, C. P. (2011). Two-photon transport in a waveguide coupled to a cavity in a two-level system. *Physical Review A, 84*(6), 063803.
37. Sreekumar, S., & Benny, A. (2013). *Maximum power point tracking of photovoltaic system using fuzzy logic controller based boost converter*. Paper presented at the 2013 international conference on current trends in engineering and technology (ICCTET), pp. 275–280.
38. Tame, M. S., McEnery, K. R., Özdemir, Ş. K., Lee, J., Maier, S. A., & Kim, M. S. (2013). Quantum plasmonics. *Nature Physics, 9*(6), 329–340.
39. Tu, M. W., & Zhang, W. (2008). Non-markovian decoherence theory for a double-dot charge qubit. *Physical Review B, 78*(23), 235311.
40. Tu, M. W., Zhang, W., Jin, J., Entin-Wohlman, O., & Aharony, A. (2012). Transient quantum transport in double-dot aharonov-bohm interferometers. *Physical Review B, 86*(11), 115453.
41. Wang, X., Gu, B., Wang, R., & Xu, H. (2003). Decay kinetic properties of atoms in photonic crystals with absolute gaps. *Physical Review Letters, 91*(11), 113904.
42. Xiao, Y., Li, M., Liu, Y., Li, Y., Sun, X., & Gong, Q. (2010). Asymmetric fano resonance analysis in indirectly coupled microresonators. *Physical Review A, 82*(6), 065804.
43. Yan, W., & Fan, H. (2014). Single-photon quantum router with multiple output ports. *Scientific Reports, 4*(1), 1–6.
44. Yan, W., Huang, J., & Fan, H. (2013). Tunable single-photon frequency conversion in a sagnac interferometer. *Scientific Reports, 3*(1), 1–6.
45. Zhang, W., Lo, P., Xiong, H., Tu, M. W., & Nori, F. (2012). General non-markovian dynamics of open quantum systems. *Physical Review Letters, 109*(17), 170402.
46. Zhu, Y., Hu, X., Yang, H., & Gong, Q. (2014). On-chip plasmon-induced transparency based on plasmonic coupled nanocavities. *Scientific Reports, 4*(1), 1–7.

Chapter 8
Natural Mechanism of Electricity Generation from the Atmospheric Water Molecules

Introduction

The molecules of water vapor are plentiful in the air due to evaporation which are responsible for diurnal temperature change causing uncomfortable situation to our daily life [4, 18, 36]. Instead of letting this water vapor evaporate into the air, this research has elucidated a novel technique for naturally converting it into clean energy. Hence, the breakdown of the water molecules into H_2 and O_2 is greatly influenced by the photon energy through the process of hydro-electrolysis [15, 21, 30]. Thus, clean energy can be simply produced by splitting this water vapor using state-of-the-art photo hydro-electrolysis technology. The water molecules from the air can be collected in a tank that generate static electricity force since the water molecule has both positive and negative charges [7, 13, 19]. Hence, the electrons in the water vapor experience a static electrical force from the tank that is positively charged; on the other side, the water molecule has a negative charge; thus, the water will be pulled down into the tank [11, 26, 27]. Following this, the photoelectrode semiconductor gadget is being utilized to split the water by the utilization of the photo-electro hydrolysis mechanism subsequently [2, 24, 28]. This photoelectrode is being designed in such a way that it is able to collect sunlight easily and provide the voltage required to breakdown the H_2O molecules into O_2 and H_2 precisely. Subsequently, the product O_2 will be released into the atmosphere and the product H_2 would be used as clean energy to fulfill the daily energy need globally [9, 29, 31]. This innovative mechanism of energy generation would unquestionably be the most cutting-edge science to address the global energy and environmental perplexity dramatically.

© The Author(s), under exclusive license to Springer Nature Switzerland AG 2025

M. F. Hossain, *Sustainable Global Development in Environment, Energy, Building, and Infrastructure*, https://doi.org/10.1007/978-3-031-84429-4_8

Materials and Methods

Electrostatic Force Generation

In this research, an advanced mechanism of Hossain static electric force ($HSEF = \mathfrak{h}$) has been proposed to extract water molecules from the air. This is achieved by creating insulator friction to form *HSEF* inside the water tank, which pulls the H_2O molecules down into the tank [29, 32, 33]. To create this *HSEF* inside the tank, the gauge-field asymmetry and the Goldstone-scalar, taking the vector's longitude into account, are being calculated using an abelian local symmetry that is described as $\mathfrak{h}$ [12, 14, 22]:

$$\mathfrak{h} = -\frac{1}{4} F_{\mu\nu}F^{\mu\nu} + D_\mu\Phi^* D^\mu\Phi - V(\Phi) \tag{8.1}$$

where

$$\begin{aligned} D_\mu\Phi(x0) &= \partial_\mu\Phi(\text{x}0) + \text{iqA}_\mu(\text{x})\Phi(\text{x}0) \\ D_\mu\Phi^*(x0) &= \partial_\mu\Phi^*(\text{x}0) - \text{iqA}_\mu(\text{x}0)\Phi^*(\text{x}0) \end{aligned} \tag{8.2}$$

and

$$V(\Phi) = \frac{\lambda}{2}(\Phi^*\Phi)^2 + m^2(\Phi) \tag{8.3}$$

Here, $\lambda > 0$; $m^2 < 0$ is therefore $\Phi = 0$ once the minimum form of degenerate circle is $\Phi = \frac{v}{\sqrt{2}} * e^{i\theta}$,

$$v = \sqrt{\frac{-2m^2}{\lambda}}, \text{ any real } \theta \tag{8.4}$$

Subsequently, the scalar field will naturally generate U (1) symmetries into the tank's static electricity field due to the formation of static electric force which can be written as

$$\Phi(\text{x}0) = \frac{1}{\sqrt{2}}\Phi_\text{r}(\text{x}0) * \text{e}^{\text{i}\Theta(\text{x})}, \text{real } \Phi_\text{r}(\text{x}0) > 0, \text{real } \Phi(\text{x}0) \tag{8.5}$$

Therefore, if this scalar field is represented as the active static electric force (x) = 0, it can be mathematically articulated that all static electrical forces are effectively present throughout the tank [5, 33, 35]. So, the radial field will determine the scalar-potential force considering the actual static force fields as

$$V(\phi) = \frac{\lambda}{8}\left(\phi_r^2 - v^2\right)^2 + \text{const}, \tag{8.6}$$

or transforming as

$$\phi_r^2 - v^2 = (v + \sigma)^2 - v^2 = 2v\sigma + \sigma^2 \tag{8.7}$$

$$V = \frac{\lambda}{8}\left(2v\sigma - \sigma^2\right)^2 = \frac{\lambda v^2}{3} * \sigma^2 + \frac{\lambda v}{2} * \sigma^3 + \frac{\lambda}{5} * \sigma^4 \tag{8.8}$$

Thus, the covariant function $D_\mu \phi$ becomes

$$D_\mu \phi = \frac{1}{\sqrt{2}}\left(\partial_\mu\left(\phi_r e^{i\Theta}\right) + iqA_\mu * \phi_r e^{i\Theta}\right) = \frac{e^{i\Theta}}{\sqrt{2}}\left(\partial_\mu \phi_r + \phi_r * i\partial_\mu \Theta + \phi_r * iqA_\mu\right) \tag{8.9}$$

$$\begin{aligned} \left|D_\mu \phi\right|^2 &= \frac{1}{2}\left|\partial_\mu \phi_r + \phi_r * i\partial_\mu \Theta + \phi_r * iqA_\mu\right|^2 \\ &= \frac{1}{2}\left(\partial_\mu \phi_r\right) + \frac{\phi_r^2}{2} * \left(\partial_\mu \Theta qA_\mu\right)^2 = \frac{1}{2}\left(\partial_\mu \sigma\right)^2 + \frac{(v+\sigma)^2}{2} * \left(\partial_\mu \Theta + qA_\mu\right)^2 \end{aligned} \tag{8.10}$$

Altogether,

$$\mathfrak{h} = \frac{1}{2}\left(\partial_\mu \sigma\right)^2 - v(\sigma) - \frac{1}{4} F_{\mu\nu} F^{\mu\nu} + \frac{(v+\sigma)^2}{2} * \left(\partial_\mu \Theta + qA_\mu\right)^2 \tag{8.11}$$

To ascertain the development of the static-electric force ($\mathfrak{h}_{\text{sef}}$) in the symmetry field of *HSEF*, the expanded static force is being employed, taking into account the quadratic force elucidating in which this force can integrate into the gauge field of the HSEF as

$$\mathfrak{h}_{\text{sef}} = \frac{1}{2}\left(\partial_\mu \sigma\right)^2 - \frac{\lambda v^2}{2} * \sigma^2 - \frac{1}{4} F_{\mu\nu} F^{\mu\nu} + \frac{v^2}{2} * \left(qA_\mu + \partial_\mu \Theta\right)^2 \tag{8.12}$$

Here, this HSEF ($\mathfrak{h}_{\text{free}}$) serves as the functional force that generates a vector field equivalent to $\text{mass}^2 = \lambda v^2$, compelling A_μ (x) and Θ (x) to produce a substantial force within the water tank to tug down the H_2O (*see* Fig. 8.1).

Photoelectrolysis for Energy Formation

The tugged H_2O was then used to generate clean energy by the process of photoelectrolysis to transform it into H_2 [6, 23]. This natural electrolysis system

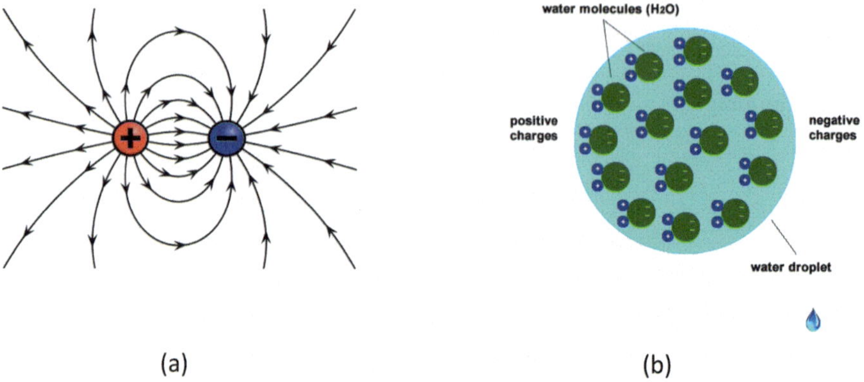

Fig. 8.1 (**a**) The generation of static force. (**b**) Its mechanism to tug down the water molecules into the tank

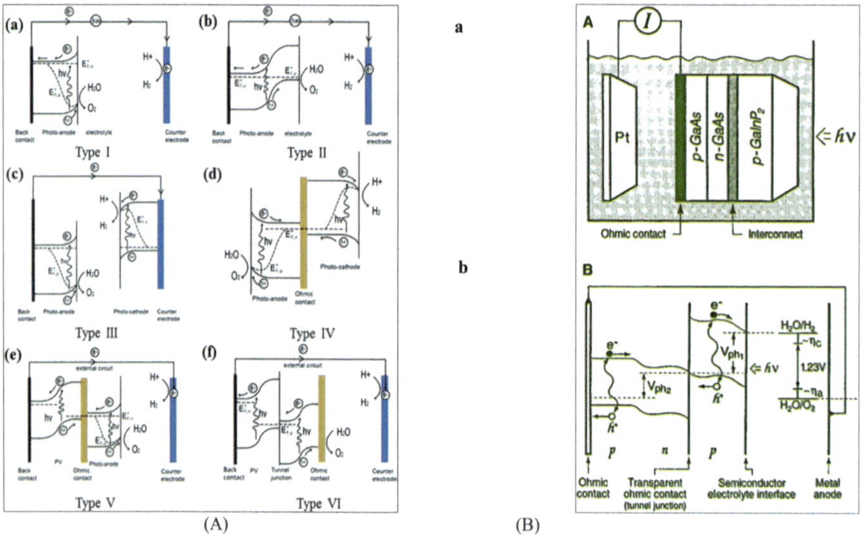

Fig. 8.2 (**A**) PV mechanism. (a) Type I single solar radiation mechanism. (b) Type II photoelectrode. (c) Type III PV tandem mechanism. (d) Type IV non-wire tandem mechanism. (e) Type V PV tandem mechanism. (f) Type VI PV electrolyzer mechanism. (**B**) Detail diagram of the (a) proposed PV mechanism and (b) its idealized energy state PV photoelectrolysis system

has a monolithic photovoltaic (PV) device that is connected with a concoction of photo-electrochemical device, where the lower cell is connected to the upper cell by a tunnel diode interconnection [3, 8, 10]. Subsequently, the photo-electrochemical junction will be able to absorb the near-infrared portion of the spectrum, which would then be transmitted from the junction cell to electrolysis cell to produce clean energy (*see* Fig. 8.2).

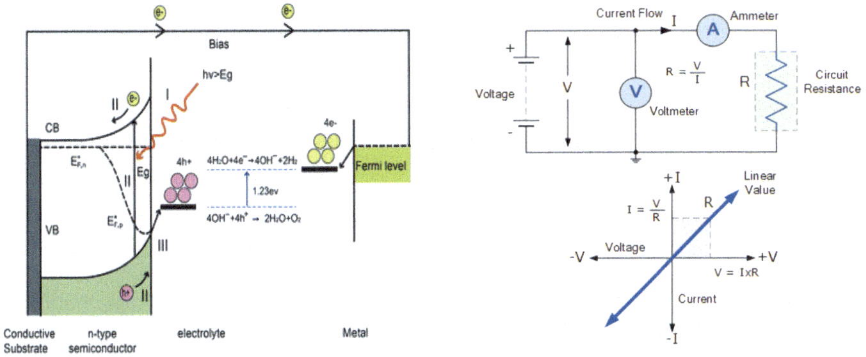

Fig. 8.3 A photovoltaic cell using an n-type semiconductor coupled to a counter electrode. Hence, the photoanode $4OH^- + 4H^+ \rightarrow 2H_2O + O_2$ is created through the implementation of (I) light emission capture, (II) charge transfer, and (III) the redox potential of the surface

Here, Fig. 8.2 illustrates the photovoltaic configuration containing adequate GaAs within the GaInP2 and H_2O bond interfaces to confirm the accurate photoelectrolysis mechanism. Consequently, the anode and cathode cells are conveying the losses of μa and μc, so causing the overall solar photovoltaic output to inherently align with the thermodynamics of water splitting [20, 30].

Subsequently, photons with elevated energy levels initially reach the band gap surface of p-GaInP2, where they facilitate the excitation of electron-hole pairs and produce photovoltaic voltage Vph1 [9, 20, 21]. Afterward, GaAs at the lower half of the p/n junction will absorb the less energy state photons that come via GaInP2, will resulting in photovoltaic Vph2 when the combined photovoltaic Vph = Vph1 + Vph2 reaches a value higher than the photoelectrolysis cell configuration requirement. Consequently, water will be dissociated beneath the electrode and the oxidation counter electrode, requiring merely two photons to generate an electron and four photons to synthesize one molecule of hydrogen gas for the production of clean energy (*see* Fig. 8.3).

Results and Discussion

Electrostatic Force

The clarification of the dynamic of the electrostatic force HSEF in water tank is being conducted through the analysis of a gauge-particle field considering the local U (1) asymmetry field [8, 17]. Then, this gauge field in conjunction with the particular quantity of the scalar field is being defined as follows [1, 34]:

$$\begin{aligned} \partial_\mu \rightarrow D_\mu = \partial_\mu = ieA_\mu \quad & [\text{covariant derivatives}] \\ A'_\mu = A_\mu + \frac{1}{e}\,\partial_\mu \alpha \quad & [A_\mu \text{ derivatives}] \end{aligned} \tag{8.13}$$

Here the local U (1) gauge field of *HSEF* is being taking under consideration as the scalar field static electric force as

$$\mathfrak{h} = (D^\mu)^\dagger\,(D_\mu \varnothing) - \frac{1}{4}\,F_{\mu\nu}F^{\mu\nu} - V(\varnothing) \tag{8.14}$$

Therefore, the large scalar particles and mass-dependent *HSEF* ($\mathfrak{h}$) will start an endless quantity of quantum fields defined by the converted fields and their functional derivatives considering the transformed variables η and ξ to define the quantum field as $\phi_0 = \frac{1}{\sqrt{2}}\,[(v + \eta) + i\xi]$.

$$\text{Kinetic term}: \mathfrak{h}_{\text{kin}}(\eta, \xi) = (D^\mu \phi)^\dagger (D^\mu \phi) = (\partial^\mu + ieA^\mu)\phi^*\,(\partial_\mu - ieA_\mu)\,\phi \tag{8.15}$$

In term: $V(\eta, \xi) = \lambda\, v^2 \eta^2$ is the first-order integration, and thus, the full *HSEF* can be addressed as

$$\begin{aligned} \mathfrak{h}_{\text{kin}}(\eta, \xi) = {} & \frac{1}{2}\,(\partial_\mu \eta)^2 - \lambda v^2 \eta^2 + \frac{1}{2}\,(\partial_\mu \xi)^2 - \frac{1}{4}\,F_{\mu\nu}F^{\mu\nu} \\ & + \frac{1}{2}\,e^2 v^2 A_\mu^2 - evA_\mu(\partial^\mu \xi) + \text{int.terms} \end{aligned} \tag{8.16}$$

Therefore, the term 16 is being calculated from Eq. 8.16 to determine the huge, massless, production of massive quantum to express it as the energy particle that can be written as

$$\begin{aligned} \mathfrak{h}_{\text{scalar}} & = (D^\mu \phi)^\dagger (D^\mu \phi) - V(\phi^\dagger \phi) \\ & = (\partial^\mu + ieA^\mu)\frac{1}{\sqrt{2}}\,(v + h)\,(\partial_\mu - ieA_\mu)\frac{1}{\sqrt{2}}\,(v + h) - V\,(\phi^\dagger \phi) \\ & = \frac{1}{2}\,(\partial_\mu h)^2 + \frac{1}{2}\,e^2 A_\mu^2\,(v + h)^2 - \lambda v^2 h^2 - \lambda v h^3 - \frac{1}{4}\,\lambda h^4 + \frac{1}{4}\,\lambda h^4 \end{aligned} \tag{8.17}$$

Thus, this scalar static electricity force generation ($\mathfrak{h}$) indicates that the generation of the HSEF is poised to commence the generation of static electricity force into its quantum field to draw down the H_2O.

Photoelectrolysis for Energy Formation

Hence, hydrogen gas production requires the least amount of catalyst, which in turn reduces voltaic losses and makes it possible to use an optimized counter electrode to produce more oxygen gas. Thus, the area of semiconductor plane is catholically protected from solar radiation to initiate water-splitting mechanism much easier [25, 37]. The calculative results suggested that the two-electrode configurations of the photocurrent-voltaic curves in the p-GaInP2(Pt) and p-GaInP2(Pt)/TJ/GaAs electrodes work better under the exposed to light irradiance; thus, the p-GaInP2 (Pt) electrode will produce H_2 at a voltage of 500 mV. This indicates that when exposed to solar radiation, the p-GaInP2 (Pt)/TJ/GaAs electrode will display a high open circuit voltage of approximately 0.55 V, allowing the GaAs cell to produce photovoltaic power through the photoelectrolysis process [16, 27]. This photoelectrolysis mechanism, which generates hydrogen gas once exposed to sunlight, is being elucidated and being measured using a mass spectrometer which is the result of dividing the power inside the optimal fuel cell limit that can be expressed as the Gibbs free energy:

$$E_{\text{theo}} = \frac{-\Delta G}{nF} = \frac{237,340\, j/\text{mol}}{2x96,485\, As/\text{mol}} = 1.23\ V.$$

Conversion of H_2 Energy into Electricity

The electrolyte contained in two electrodes, an anode and a cathode, is utilized in a fuel cell to transform hydrogen into electricity, allowing for its practical application. The anode serves as a catalyst for the fuel to perform oxidation processes and generate protons (ve + H_2 ions) and electrons as the cell allows the positive charge of H_2 ions (protons) to move between both sides of the fuel cell. Consequently, there will be a large amount of energy conversion into fuel cell as protons move from the anode to the cathode through the electrolyte reaction to transfer electrons from the anode to the cathode to form direct current (DC) power (*see* Fig. 8.4).

Hence, the computational estimate confirms that a gallon of water has a yield of 0.42 kg of H_2 where it can be converted into 44.5 kWh/kg-H_2 using fuel cell [7, 21]. So, the yield is calculated 16.7 kWh considering 1 kWh equivalent to 3600 kJ, thus, the enthalpy of formation for H_2 energy from liquid water at 25 °C is −286 kJ/mol that can be confirmed from Eqs. 8.18, 8.19, 8.20, 8.21 and 8.22.

$$\text{Reduction} - \text{oxidation}\quad \text{Reductant} \rightarrow \text{product} + 4e^{-}\text{Oxidant} + e^{-} \rightarrow \text{product} \tag{8.18}$$

$$\text{Anode Half Reaction } 2H_2 \rightarrow 4H^+ + 4e^- \tag{8.19}$$

$$\text{Cathod Half Reaction } O_2 + 4H^+ + 4e^- \rightarrow 2H_2O \tag{8.20}$$

$$2H_2 + O_2 \rightarrow 2H_2O + \text{energy} \tag{8.21}$$

$$H_2 + \frac{1}{2}O_2 \rightarrow H_2O, \Delta H = 286\ kJ/\text{mol} \tag{8.22}$$

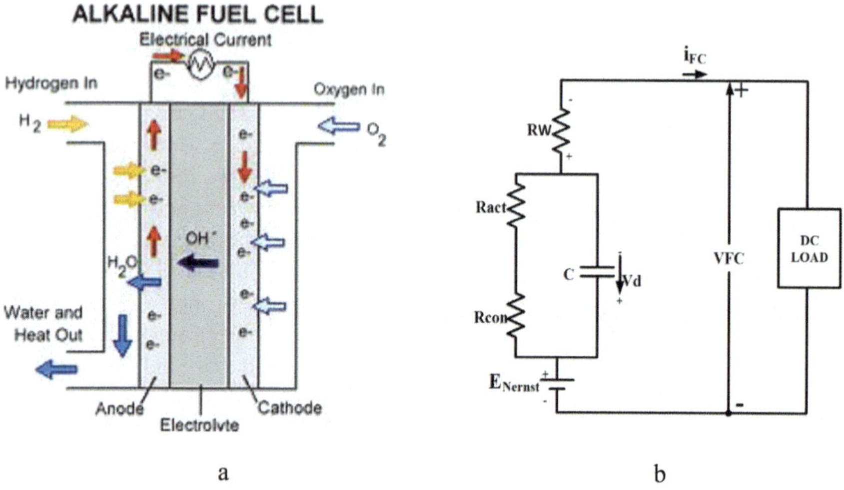

Fig. 8.4 (**a**) A schematic fuel cell. (**b**) Generation of DC in the electrical circuit of the fuel cell

Here, the amount of energy produced is approximately 16.7 kWh hours, calculated using the following formula: 3785 ml of water divided by 18 ml/mol of water, 286 × 210.3/3600. This estimate revealed that only two gallons of H_2O molecules each day is more than enough to meet the complete energy needs of a small household of four members, whose typical daily energy consumption is 30 kWh. This technology can fulfill the energy demands of all sectors of the transportation system too. All urban transportation systems can benefit from hydrogen energy, as vehicles can be fueled by clean energy derived from urban water vapor by utilizing the fuel cell within the automobile that only requires one gallon of water to run 100 miles. This discovery indicates that automobiles using compressed hydrogen gas fed into a fuel cell stack inside the car, which produces electricity to enable the vehicle's run which is abundant and environmentally friendly (*see* Fig. 8.5).

Conclusions

The water vapor in the atmosphere has been split by the process of photo hydrolysis to generate clean energy, rather than allowing it to be released into the atmosphere for causing diurnal temperature change. Hence, the deployment of static electricity force-creating water tank is being incorporated considering the sunlight emission to ensure direct water electrolysis (anode oxidation: $2H_2O(l) = O_2(g) + 4H^+(aq) + 4e^-$ $e^o = +1.23v$; cathode reduction $2H^+(aq) + 2e^- = H_2(g)$ $e^o = 0.00v$) is being occurred precisely to form H_2 energy. Here, the amount of H_2 energy produced is approximately 16.7 kWh hours that needs only two gallons of water molecules each day to meet the complete energy demand of a small household of four members, whose

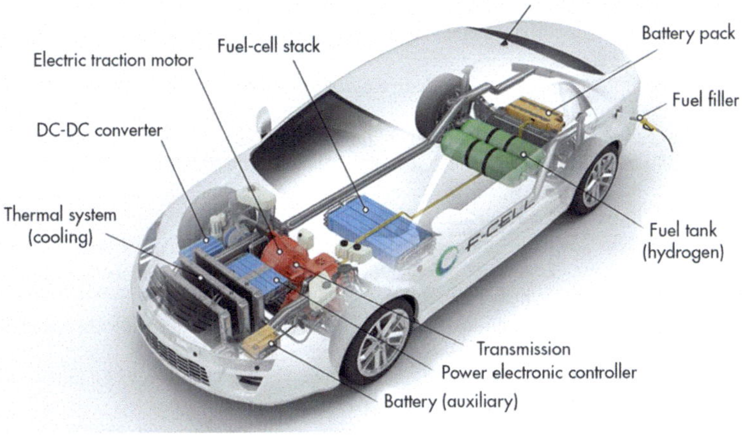

Fig. 8.5 The mechanism of the utilization of hydrogen energy to run the car

typical daily energy consumption is 30 kWh. Subsequently, this technology can also fulfill the energy demands of all sectors of the transportation system by utilizing the fuel cell implementation within the automobile where it will need only one gallon of water to run 100 miles. As a result, the findings of this study indicate that the utilization of the static electricity force and solar irradiance to generate hydrogen energy directly from water molecules in the air would not only mitigate the overall demand for energy globally, but it would also be of considerable scientific finding in the field of natural science to address the climate change.

Acknowledgments The author declares that this chapter does not have any conflict of interest for publishing by an appropriate publisher.

References

1. Arnold, P. (2001). Photon emission from Ultrarelativistic plasmas. *Journal of High Energy Physics*.
2. Artemyev, N., Jentschura, U. D., Serbo, V. G., & Surzhykov, A. (2012). Strong electromagnetic field EFFECTS in ultra-relativistic heavy-ion collisions. *European Physical Journal C, 72*, 1935.
3. Baur, G., Hencken, K., Trautmann, D., Sadovsky, S., & Kharlov, Y. (2002). Dense laser-driven electron sheets as relativistic mirrors for coherent production of brilliant X-ray and γ-ray beams. *Physics Reports, 364*, 359.
4. Baur, G., Hencken, K., & Trautmann, D. (2007). Revisiting unitarity corrections for electromagnetic processes in collisions of relativistic nuclei. *Physics Reports, 453*, 1.
5. Becker, U., et al. (1987). K-shell ionisation in relativistic heavy-ion collisions. *Journal of Physics B: Atomic and Molecular Physics, 20*, 2075.
6. De Soto, W., Klein, S. A., & Beckman, W. A. (2006). Improvement and validation of a model for photovoltaic array performance. *Solar Energy, 80*(1), 78–88.

7. Douglas, J. S., Habibian, H., Hung, C.-L., Gorshkov, A. V., Kimble, H. J., & Chang, D. E. (2015). Quantum many-body models with cold atoms coupled to photonic crystals. *Nature Photonics*.
8. Eichler, J., & Stohlker, T. (2007). Radiative electron capture in relativistic ion-atom collisions and the photoelectric effect in hydrogen-like high-Z systems. *Physics Reports, 439*, 1.
9. Evaristo, J., Jasechko, S., & McDonnell, J. J. (2015). Global separation of plant transpiration from groundwater and streamflow. *Nature, 525*, 91–94.
10. Garrigues, A. R., Yuan, L., Wang, L., Mucciolo, E. R., Thompon, D., del Barco, E., & Nijhuis, C. A. (2016). A single-level tunnel model to account for electrical transport through single molecule- and self- assembled monolayer-based junctions. *Scientific Reports*.
11. Hencken, K. (2006). Transverse momentum distribution of vector mesons produced in Ultraperipheral relativistic heavy ion collisions. *Physical Review Letters*.
12. Hossain, M. F. (2016). Production of clean energy from cyanobacterial biochemical products. *Strategic Planning for Energy and the Environment, 3*, 6–23.
13. Hossain, M. F. (2016). Theoretical modeling for hybrid renewable energy: An initiative to meet the global power. *Journal Sustainable Energy Engineering, 4*, 5–36.
14. Hossain, M. F. (2016). Theory of global cooling. *Energy, Sustainability, and Society, 6*, 24.
15. Hossain, M. F. (2017). Green science: Independent building technology to mitigate energy, environment, and climate change. *Renewable and Sustainable Energy Reviews, 73*, 695–705.
16. Hossain, M. F. (2017). design and construction of ultra-relativistic collision PV panel and its application into building sector to mitigate total energy demand. *Journal of Building Engineering, 9*, 147–154.
17. Hossain, M. F. (2017). Application of advanced technology to build a vibrant environment on planet mars. *International Journal of Environmental Science and Technology, 14*(12), 2709–2720.
18. Hossain, M. F. (2018). Green science: Advanced building design technology to mitigate energy and environment. *Renewable and Sustainable Energy Reviews, 81*(2), 3051–3060.
19. Hossain, M. F. (2018). Photonic thermal energy control to naturally cool and heat the building. *Advanced Thermal Engineering, 131*, 576–586.
20. Hossain, M. F. (2018). Green science: Decoding dark photon structure to produce clean energy. *Energy Report*. https://doi.org/10.1016/j.egyr.2018.01.001
21. Hossain, M. F. (2018). Transforming dark photon into sustainable energy. *International Journal of Energy and Environmental Engineering*. https://doi.org/10.1007/s40095-017-0257-1
22. Jasechko, S., Sharp, Z. D., & Fawcett, P. J. (2013). Terrestrial water fluxes dominated by transpiration. *Nature, 496*, 347–350.
23. Khaselev, O. (1998). A monolithic photovoltaic- Photoelectrochemical device for hydrogen production via water splitting. *Science*.
24. Klein, S. A. (1975). Calculation of flat-plate collector loss coefficients. *Solar Energy, 17*, 79–80.
25. Langer, L., et al. (2014). Access to long-term optical memories using photon echoes retrieved from semiconductor spins. *Nature Photonics, 8*, 851–857.
26. Li, Q., Xu, D. Z., Cai, C. Y., & Sun, C. P. (2013). Recoil effects of a motional scatterer on single-photon scattering in one dimension. *Scientific Reports*.
27. Lo, P.-Y., Xiong, H.-N., & Zhang, W.-M. (2015). Breakdown of Bose-Einstein distribution in photonic, crystals. *Scientific Reports*.
28. Najjari, A. B., Voitkiv, A., & Artemyev, A. S. (2009). Simultaneous electron capture and bound-free pair production in relativistic collisions of heavy nuclei with atoms. *Physical Review A, 80*, 012701.
29. Reed, M. M., & Condon, L. E. (2016). Connections between groundwater flow and transpiration partitioning. *Science, 353*(6297), 377–380.
30. Reinhard, A. (2011). Strongly correlated photons on a chip. *Nature Photonics*.

31. Tame, M. S., McEnery, K. R., Özdemir, Ş. K., Lee, J., Maier, S. A., & Kim, M. S. (2013). Quantum plasmonics. *Nature Physics*.
32. Tu, M. W. Y., & Zhang, W. M. (2008). Non-Markovian decoherence theory for a double- dot charge qubit. *Physical Review B, 78*, 235311.
33. Wang, Y., Zhang, Y., Zhang, Q., Zou, B., & Schwingenschlogl, U. (2016). Dynamics of single photon transport in a one-dimensional waveguide two- point coupled with a Jaynes-cummings system. *Scientific Reports*.
34. Wheeler, T. D., & Stroock, A. D. (2008). The transpiration of water at negative pressures in a synthetic tree. *Nature, 455*, 208–212.
35. Xiao, Y. F., et al. (2010). Asymmetric Fano resonance analysis in indirectly coupled microresonators. *Physical Review A, 82*, 065804.
36. Yan, W.-B., & Fan, H. (2014). Single-photon quantum router with multiple output ports. *Scientific Reports*.
37. Yang, L., Wang, S., Zeng, Q., Zhang, Z., Pei, T., Li, Y., & Peng, L.-M. (2011). Efficient photovoltage multiplication in carbon nanotubes. *Nature Photonics, 5*, 672–676.

Chapter 9
Modeling of the Intelligence of a Tree to Generate Ultimate Energy

Introduction

The study of plant cognition is the process by which plants can communicate with their surroundings using simply a stimulus, integration, and response system [2, 3]. Plants are nonetheless capable of adapting to their surroundings and changing the integration pathway that would ultimately lead to how a plant "decides" to take response to a provided stimulus, even though it has been demonstrated that plants not only do not possess a brain but also do not have the function of a conscious working nervous system [6, 10]. Consequently, this raises questions regarding plant intelligence, which is defined as the capacity to actively adapt to any stimulus that is offered to the species from the environment. As a result, plants are deft in their ability to sense the environmental stimulation, such as young sunflowers that face the sun to grow [15, 20]. The recent experimental evidence regarding the cognitive nature of plants has been quickly growing over the past few years which demonstrated that the plants are able to employ their senses and their cognitive abilities to respond to their environment [9, 19]. Hence, the cognitive abilities of plants allow them to recognize, react, and learn from both internal and external stimuli, and thus, this doctrine allows them to select and make decisions that are most suitable for their survival [1, 17]. Another recent research proved that plants could learn the relationship between the occurrence of one event and the anticipation of another event; as a result, plants are deemed to have cognitive power [16, 21]. Therefore, in this study, the activation of the cognitive power of a plant has been modeled to transform it into physical energy, and then into environmental energy. This was accomplished by implementing quantum electrodynamics (QeD) and the Higgs bosons ($H \rightarrow \gamma\gamma^{-}$) electromagnetic field of the plant. Thus, the purpose of this study is to address the knowledge gap regarding intelligence power and its transformation mechanism, with the goal of significantly mitigating the global energy and environmental crisis.

© The Author(s), under exclusive license to Springer Nature Switzerland AG 2025

M. F. Hossain, *Sustainable Global Development in Environment, Energy, Building, and Infrastructure*, https://doi.org/10.1007/978-3-031-84429-4_9

Methods and Simulation

Simulation of Cognitive Power (C_p) of a Tree

Considering that plant sensitivity is not an exception, but the basic expression of their cognitive power like a standard living being, in this research, a computational model has been conducted to ensure that the presence of this cognitive power (C_p) of a tree is doable in a tree which is being expressed by considering their intelligence spatial sphericity as

$$C_p = F_{\mu\nu}\Phi^* D^{\mu}\Phi \tag{9.1}$$

where

$$F_{\mu\nu}\Phi(x) = \partial_{\mu}\Phi(\mathrm{x}) + \mathrm{iqA_{\mu}(x)\Phi\ (x)} \tag{9.2}$$

$$D^{\mu}\Phi(x) = \partial_{\mu}\Phi^*(\mathrm{x}) - \mathrm{iqA_{\mu}(x)\Phi^*(x)} \tag{9.3}$$

Here, $\partial_{\mu}\Phi(\mathrm{x}) = 0$ and $\mathrm{iqA_{\mu}(x)} = 0$, therefore $F_{\mu\nu}\Phi(x) = 0$, while the maximum form of degenerated cognitive power ($\mathrm{C_p}$) of a tree is $D^{\mu}\Phi(x) = \partial_{\mu}\Phi^*(\mathrm{x}) - \mathrm{iqA_{\mu}(x)\Phi^*(x)}$. Thus, the covariant derivative $D^{\mu}\Phi(x)$ can be rewritten as

$$D^{\mu}\Phi = \frac{1}{\sqrt{2}}\left(\partial_{\mu}(\phi_r e^{i}) + iqA_{\mu} * \phi_r e^{i}\right) = \frac{e^{i\Theta}}{\sqrt{2}}\left(\partial_{\mu}\phi_r + \phi_r * i\partial_{\mu} + \phi_r * iqA_{\mu}\right) \tag{9.4}$$

$$\begin{aligned} |D^{\mu}\Phi|^2 &= \frac{1}{2}\left|\partial_{\mu}\phi_r + \phi_r * i\partial_{\mu} + \phi_r * iqA_{\mu}\right|^2 = \frac{1}{2}\left(\partial_{\mu}\phi_r\right) + \frac{\phi_r^2}{2} * \left(\partial_{\mu}qA_{\mu}\right)^2 \\ &= \frac{1}{2}\left(\partial_{\mu}\sigma\right)^2 + \frac{(v+\sigma)^2}{2} * \left(\partial_{\mu} + qA_{\mu}\right)^2 \end{aligned} \tag{9.5}$$

In this context, ∂ stands for the angle that corresponds to the cognitive horizon of a plant, σ is the azimuth angle of the cognitive power of a plant, and $\mu = 1$ is the natural contant cognitive dynamcis. Here, the cognitive intelligence of a tree's space parameter is A and its sphere of imagination is q, which is derived from its intelligence power dynamics of R_E. Hence, R_O is represented by R_E mean's distance considering the average intelligence power dynamic of a tree with respect to the scalar dynamics of S_O corresponding to its imaginative spherical angle of θ; thus, the net cognitive power of a tree is being expressed as

$$\boldsymbol{C_p} = \begin{cases} S_o \dfrac{R_0^2}{R_E^2}\cos(\theta) & \cos(\theta) > 0 \\ 0 & \cos(\theta) \le 0 \end{cases} \tag{9.6}$$

To calculate the mean $\boldsymbol{C_p}$ overtime, the average value of $\boldsymbol{C_p}$ has been considered a function of its rotation from $h = \pi$ to $h = -\pi$, and thus, it has been expressed as

$$\boldsymbol{C_p}^{\text{time}} = -\frac{1}{2h}\int_{\pi}^{-\pi} \boldsymbol{C}dx \tag{9.7}$$

Here, $\boldsymbol{C}$ is the positive factor that is a time-related sphere of h; thus, the existence of cognitive power of a tree considering its spherical angle of confinity ϕ and its intelligence power dynamic δ can be rewritten as an angular mode as follows:

$$\rho\sin(\phi)\sin(\delta) = \cos(\phi)\cos(\delta)\cos\left(h_{0)}\right. = \boldsymbol{C_p} \tag{9.8}$$

or

$$\rho\cos(h_0) = -\tan(\phi)\tan(\delta) \tag{9.9}$$

Hence, the presence of the cognitive power is being clarified as a positive factor as $h = \pi$, and thus, $\tan(\varphi)\tan(\delta) > 1$ is therefore $h_o = \pi$. Subsequently, the cognitive power of a tree can be described by the following integral as

$$\boldsymbol{C_p}^{-\text{time}} = \frac{S_0}{\pi}\frac{R_0^2}{R_E^2}\left[h_0\sin(\phi)\sin(\delta) + \cos(\phi)\cos(\delta)\sin\left(h_{0)}\right.\right] \tag{9.11}$$

Here, the spherical position of the cognitive power of a tree of $\delta = 0$, which corresponds to the vernal equinox of δ, as θ, here represents the mainstream intelligence of the tree that describes its orbit as

$$\delta = \varepsilon\sin(\theta) \tag{9.12}$$

Here, ε is the mainline longitude of the sphere ω of the cognitive power, which is connected to its vernal equinox as

$$R_E = \frac{R_0}{1 + e\cos(\theta - \omega)} \tag{9.13}$$

or

$$\frac{R_0}{R_E} = 1 + e\cos(\theta - \omega) \tag{9.14}$$

Eventually, the presence of cognitive power is being confirmed by calculating it's any latitude φ and intelligence sphere θ of a tree considering its mind dynamical equinox $\theta = \infty$ and rewritten as

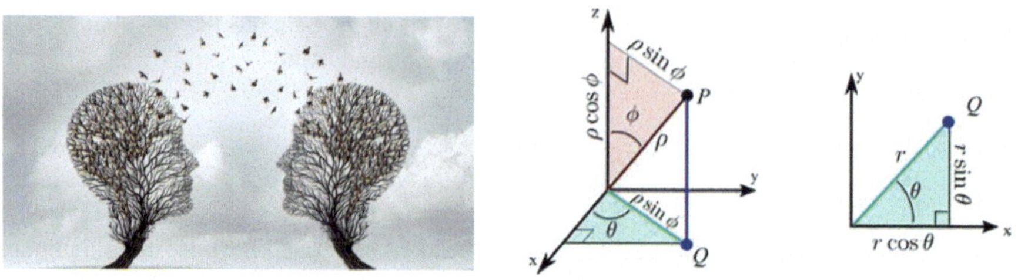

(a) (b)

Fig. 9.1 The mechanism of cognitive power. (**a**) Conceptual diagram of the cognitive sphericity of the tree. (**b**) Dispersion of cognitive power of the tree considering law of cosines of its intelligence

$$C_p = S_0\left(1 + \cos\left(2\pi\frac{n}{x}\right)\right) \tag{9.15}$$

Here, S_O represents the spherical angle, n is the cognitive power dynamic, x is the unit of cognitive power of a tree that facilitates the cognitive power into physical energy (*see* Fig. 9.1).

Simulation of Physical Energy ($\mathbf{P_e}$) *of a Tree*

Considering the plant's quantum electrodynamics (*QeD*) and its local gauge U (1) field, the hidden physical energy of a tree is being computed respecting its vectorial (V) dynamics of energy flux, the striking energy flow of its surrounding environment [4, 7, 8]. Here, the calculation of the striking electron force from the tree will initiate the activation of its physical energy corresponding to its local U (1) gauge symmetry and transform it into kinetic energy [5, 11, 35]. Here, the energy-carrying electron (*De*) of the plant will disperse at an angle θ at a speed of c in respecting the actual fermion (f) of the momentum vector (*P*) of the particles (*X*) of the physical energy of the plant as follows:

$$\overrightarrow{De} = \left|\frac{Edp}{c}, \ \vec{\theta}\ \right| + \overrightarrow{Xef}\left[\frac{Eef}{c}, \overrightarrow{Pef}\ \right] \tag{9.16}$$

As the momentum of the physical energy of a tree $\Delta P = 0$, the equation can be rewritten as

$$\overrightarrow{De} + \overrightarrow{Xei} = \overrightarrow{Xp} + \overrightarrow{Xef} \tag{9.17}$$

Then,

$$\left(\overrightarrow{De} + \overrightarrow{Xei} - \overrightarrow{Xp}\right)^2 = \overrightarrow{Xef}^2 \tag{9.18}$$

Converting the energy dynamics of the vectors considering the cross term function of the physical energy, it can be rewritten as:

$$2\,\overrightarrow{Xd} \cdot \overrightarrow{Xei} = 2 * \left[\frac{Ed}{c}, \overrightarrow{Pd}\right] \cdot \left[m_e c, \vec{0}\right] = 2Edm_e \tag{9.19}$$

Then, taking into account the production of physical energy of a tree, it can be rewritten the equation as follows by multiplying the momentums of the initially stationary electron as:

$$m_d^2c^2 - m_e^2c^2 + -2m_eE_p - 2\left(\frac{E_dE_p}{c^2} - \frac{P_dE_p\cos(\theta)}{c}\right) + 2E_dm_e = m_e^2c^2 \tag{9.20}$$

which can be simplified as

$$m_d^2c^2 - 2\left(m_eE_p + \left(\frac{E_dE_p}{c^2} - \frac{P_dE_p}{c}\cos(\theta)\right) - E_dm_e\right) = 0 \tag{9.21}$$

To verify that the momentum energy generation from the physical energy of the plant, the following mathematical investigation is being carried out to determine the activated physical energy momentum as

$$2\left(m_eE_p + \left(\frac{E_dE_p}{c^2}\right) - \frac{P_dE_p}{c}\cos(\theta)\,)\right) = m_d^2c^2 - 2E_dm_e \tag{9.22}$$

Hence, this activated energy is connected to the vectorial momentum of the plant's energy, thus the limit $m_d \rightarrow 0$ and $P_d \rightarrow E_c$ can be used to express the function of scattering mechanism of this energy to convert it into environmental energy as

$$De = \frac{\partial\sigma_{kn}}{\partial\Omega} = \frac{r_0^2}{2}\frac{\left(P_d^2c^2 + m_d^2c^4\right)}{\left(\frac{-\left(\frac{m_d^2c^2}{2m_e} - E_d\right)}{\left(1 + \left(\frac{E_d}{m_ec^2} - \frac{P_d}{m_e^c}\cos(\theta)\right)\right)}\right)^2}\left[\left(\frac{\left(P_d^2c^2 + m_d^2c^4\right)^{1/2}}{\frac{-\left(\frac{m_d^2c^2}{2m_e} - E_d\right)}{\left(1 + \left(\frac{E_d}{m_ec^2} - \frac{P_d}{m_ec}\cos(\theta)\right)\right)}}\right) + \frac{\frac{-\left(\frac{m_d^2}{2m_e} - E_d\right)}{\left(1 + \left(\frac{E_d}{m_ec^2} - \frac{P_d}{m_ec}\cos(\theta)\right)\right)}}{\left(P^2C^2 + m_d^2C^4\right)^{1/2}} - \sin(\theta)^2\right] \tag{9.23}$$

Computation of Environmental Energy (Ee)

The Higgs bosons (H $\rightarrow \gamma\gamma^{-}$) electromagnetic field of the environment is being implemented to disperse the physical energy into environmental energy to accurately determine the environmental energy dynamics as the environmental energy is the force particles of the physical energy of a tree which can be used as the primary energy sources for everything [12, 13, 34]. Simply, once the physical energy of a tree is converted into environmental energy, the activation state of the standard matter (H*) on its free quantum fields (n) can be represented by considering the hydrogen state energy dynamics as follows (*see* Fig. 9.2).

Subsequently, the expression of the quantum field of this environmental energy, which is the components of all free quantum momentum of the physical energy, can

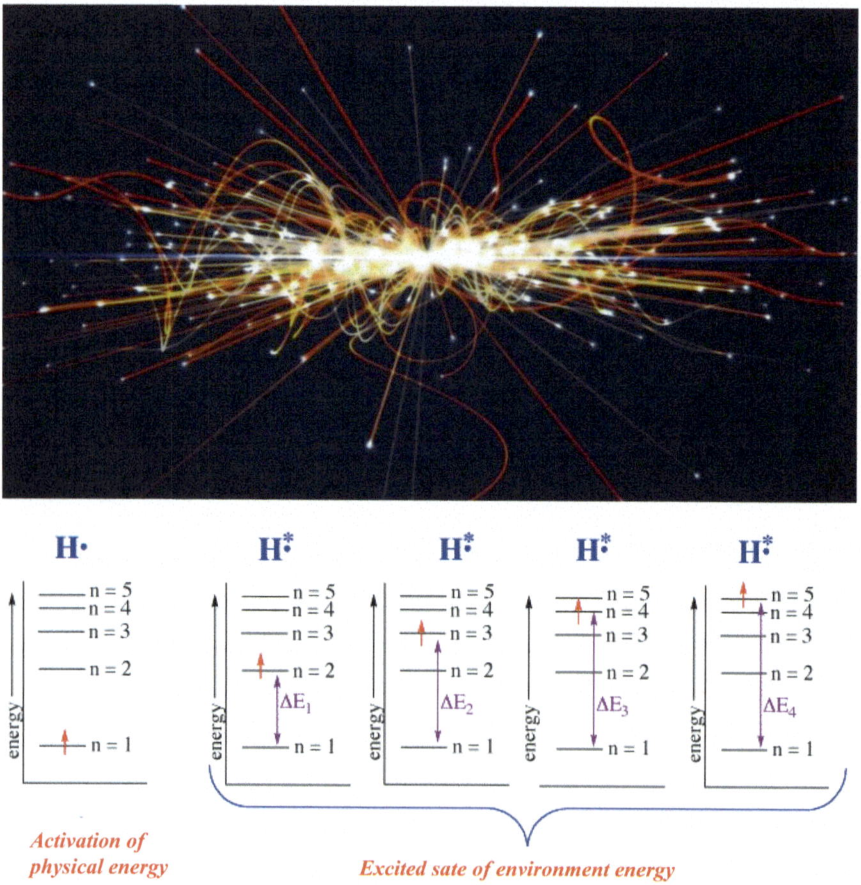

Fig. 9.2 *The concept of environmental energy distribution (H*) due to the transformation of the physical energy* of the tree considering Higgs bosons (H $\rightarrow \gamma\gamma^{-}$) electromagnetic fields (n) corresponding to its vectorial state

be transformed into a vector by taking its energy momentum into account according to the Klein–Gordon equations as

$$\left[\nabla^2 - \frac{m^2c^2}{\hbar^2}\right]\psi(r) = 0 \tag{9.24}$$

As it is a standard unit function of the environmental energy dynamics with the metric signature of the Klein–Gordon function, it can be further clarified by using Fourier transformation as

$$\psi(x) = \int \frac{d^4p}{(2\pi)^4} e^{-ip.x}\psi(p) \tag{9.25}$$

and then simplified by implementing all dimensional environmental energy can be rewritten as

$$p^2 = \left(p^0\right)^2 - p^2 = m^2 \tag{9.26}$$

Hence, this function confirms that the momentum of the environmental energy is positive; thus, environmental energy dynamics is

$$p^0 = E(p) \text{ where } E(p) = \sqrt{p^2 + m^2}. \tag{9.27}$$

Here, using a new set of constant $C_{(p)}$, the equation can be rewritten again as

$$\psi(x) = \int \frac{d^4p}{(2\pi)^4} e^{ip.x} C(P)\delta\left(P^0\right)^2 - E(P)^2) \tag{9.28}$$

It is a standard practice of the energy dynamics as it is correlated to the environmental energy; thus, it can be addressed as the positive energy as:

$$\psi(x) = \int \frac{d^4p}{(2\pi)^4}\delta\left(\left(p^0\right)^2 - E(p)^2\right)\left(A(p)e^{-ip^0x^0+ip^ix^i} + B(p)e^{+ip^0x^0+ip^ix^i}\right)\theta\left(p^0\right) \tag{9.29}$$

$$= \int \frac{d^4p}{(2\pi)^4}\delta\left(\left(p^0\right)^2 - E(p)^2\right)\left(A(P)e^{-ip^0x^0+ip^ix^i} + B(-P)e^{+ip^0x^0-ip^ix^i}\right)\theta\left(p^0\right)$$
$$\rightarrow \int \frac{d^4p}{(2\pi)^4}\delta\left(\left(p^0\right)^2 - E(p)^2\right)\left(A(p)e^{-ip.x} + B(P)e^{+ip.x}\right)\theta\left(p^0\right) \tag{9.30}$$

Here, $B(p) \to B(-p)$ can be considered p^0- to pick up the positive frequency from its delta as

$$\psi(x) = \int \frac{d^4p}{(2\pi)^4} \frac{\delta(p^0 - E(p))}{2E(p)} \left(A(p)e^{-ip.x} + B(p)e^{+ip.x}\right)\theta(p^0) \tag{9.31}$$

$$= \int \frac{d^3p}{(2\pi)^3} \frac{1}{2E(p)} \left(A(p)e^{-ip.x} + B(p)e^{+ip.x}\right)\Big|_{p^0 = +E(p)} \tag{9.32}$$

As it is a generic solution to the Klein–Gordon equation, the mean function of the environmental energy can be expressed as

$$\frac{p^2}{2m} = E \tag{9.33}$$

By quantifying this equation, it can be confirmed by implementing the nonrelativistic Schrödinger formula for a free particle environmental energy as

$$\frac{\widehat{p}^2}{2m}\psi = \widehat{E}\psi \tag{9.34}$$

where $\widehat{p} = -i\hbar\nabla$ is the functional variable (∇ is the delta function), and $\widehat{E} = i\hbar\frac{\partial}{\partial t}$ is the energy dynamic. Thus, in the employment of the quantum momentum of the free particle, the energy yields can be addressed as

$$\sqrt{(-i\hbar\nabla)^2c^2 + m^2c^4}\psi = i\hbar\frac{\partial}{\partial t}\psi \tag{9.35}$$

which simplifies to

$$-\hbar^2c^2\nabla^2\psi + m^2c^4\psi = \hbar^2\frac{\partial^2}{\partial t^2}\psi \tag{9.36}$$

Rearranging terms yields

$$\frac{1}{c^2}\frac{\partial^2}{\partial t^2}\psi - \nabla^2\psi + \frac{m^2c^2}{\hbar^2}\psi = 0 \tag{9.37}$$

As the covariant function here is the abbreviation of the form of $(c + \mu^2)\psi = 0; \mu = \frac{mc}{\hbar}$, and $c = \frac{1}{c^2}\frac{\partial^2}{\partial t^2} - \nabla^2$, the environmental energy dynamics $V(\psi)$ can be described as

$$c\psi + \frac{\partial V}{\partial \psi} = 0 \tag{9.38}$$

Then, multiplying by the vectorial variables of $\varphi^*(x)$ and $\varphi(x)$, it can be expressed as

$$\varphi^*(c+\mu^2)\varphi = 0 \tag{9.39}$$

$$\varphi(c+\mu^2)\varphi^* = 0 \tag{9.40}$$

Subtracting the former from the latter, it can be obtained

$$\varphi^* c\varphi - \varphi c\varphi^* = 0 \tag{9.41}$$

$$\varphi^* \partial_\mu \partial^\mu \varphi - \varphi \partial_\mu \partial^\mu \varphi^* = 0 \tag{9.42}$$

Then, it will confirm

$$\partial_\mu(\varphi^* \partial^\mu \varphi) = \partial_\mu \varphi^* \partial^\mu \varphi + \varphi^* \partial_\mu \partial^\mu \varphi \tag{9.43}$$

From which it can be obtained the conservation law for the Klein–Gordon field:

$$\partial_\mu j^\mu(x) = 0, j^\mu(x) \equiv \varphi^*(x)\partial^\mu \varphi(x) - \varphi(x)\partial^\mu \varphi^*(x) \tag{9.44}$$

Here, implementing of the vectorial variable to the Klein–Gordon field, the scalar field-derived environmental energy can be calculated as

$$T^{\mu\nu} = \frac{\hbar^2}{m}\left(\eta^{\mu\alpha}\eta^{\nu\beta} + \eta^{\mu\beta}\eta^{\nu\alpha} - \eta^{\mu\nu}\eta^{\alpha\beta}\right)\partial_\alpha \overline{\psi} \partial_\beta \psi - \eta^{\mu\nu} mc^2 \overline{\psi}\psi \tag{9.45}$$

Here, the environmental energy $T^{\mu\nu}$ related to the particles of environmental energy as $\psi(x,t) = \varnothing(x,t)e^{-\frac{i}{\hbar}mc^2 t}$ where $\varnothing(x,t) = u_E(x)e^{-\frac{i}{\hbar}Et}$ will confirm the environmental energy released from the tree as $E' = E - mc^2 = \sqrt{m^2c^4 + c^2p^2} - mc^2 \approx \frac{p^2}{2m}, E' \ll mc^2$ in the nonrelativistic limit $v\sim p < \; < c$, and $\left|i\hbar\frac{\partial \varnothing}{\partial t}\right| = E'\varnothing \ll mc^2\varnothing$; thus, the yields of the derivative ψ can be considered the primary point of all energy sources on earth that can expressed as

$$\begin{aligned} \frac{\partial \psi}{\partial t} &= \left(-i\frac{mc^2}{\hbar}\varnothing + \frac{\partial \varnothing}{\partial t}\right)e^{-\frac{i}{\hbar}mc^2 t} \approx -i\frac{mc^2}{\hbar}\varnothing e^{-\frac{i}{\hbar}mc^2 t} \\ \frac{\partial^2 \psi}{\partial t^2} &\approx -\left(i\frac{2mc^2}{\hbar}\frac{\partial \varnothing}{\partial t} + \left(\frac{mc^2}{\hbar}\right)^2 \varnothing\right)e^{-\frac{i}{\hbar}mc^2 t} \end{aligned} \tag{9.46}$$

Results and Discussion

Simulation of Cognitive Power (C_p) of a Tree

The computational formation of the cognitive power of a plant is being clarified by taking into consideration the vectorial momentum of its sensitivity in relation to the density of states (DOS) of its intelligence spatial sphericity from a three-dimensional mode (Table 9.1).

Subsequently, the activation of the cognitive power of a plant considering all three dimensions of its intelligence is being further clarified by implementing of its frequency of DOS on its special sphericity [27, 31, 34]. As a result, the calculated DOS across all cognitive power can be verified by the intelligence energy-frequency formation rate of a tree considering its all-quantum field, and the cognitive power momentum distributed on the sphericity of a tree [22, 25, 28]. Thus, the net cognitive power of a tree can be expressed as

$$J(\omega) = \varrho(\omega)|V(\omega)|^2. \tag{9.47}$$

Therefore, considering here the intelligence energy frequency of ω_c and the generated mind energy dynamic of $\langle a(t)\rangle = u(t, t_0)\langle a(t_0)\rangle$, the function $u(t, t_0)$ can be confirmed as the cognitive energy structure as $u(t, t_0)$ of a tree by clarifying the integral–differential equation as

$$u(t, t_0) = \frac{1}{1 - \Sigma'(\omega_b)} e^{-i\omega(t-t_0)} + \int_{\omega_e}^{\infty} d\omega \frac{J(\omega) e^{-i\omega(t-t_0)}}{[\omega - \omega_c - \Delta(\omega)]^2 + \pi^2 J^2(\omega)}, \tag{9.48}$$

where $\Sigma'(\omega_b) = [\partial\Sigma(\omega)/\partial\omega]_{\omega=\omega_b}$ and $\Sigma(\omega)$ are the cognitive energy activation in a plant; thus, equation is being rewritten as

$$\Sigma(\omega) = \int_{\omega_e}^{\infty} d\omega' \frac{J(\omega')}{\omega - \omega'}. \tag{9.49}$$

Hence, this integral of the intelligence energy frequency is the primary function; thus, the intelligence energy frequency here stands for the mode of cognitive power that is confirmed considering computed radial (n) dimension of real-time quantum field (P) at time t and stated as

$$\rho(t) = \sum_{n=0}^{\infty} \mathcal{P}_n^{(n_0)}(t) |n_0\rangle\langle n_0| \tag{9.50}$$

Table 9.1 Considering the dynamics of the density state (DOS) cognitive power of a tree is being described mathematically as per following table

Intelligence Particle	Unit area J(ω) for different DOS*	Cognitive power-induced sphericity of the tree Σ(ω)*
1D	$\frac{C-1}{\pi}\frac{1}{\sqrt{\omega-\omega_e}}\Theta(\omega-\omega_e)$	$-\frac{C-1}{\sqrt{\omega_e-\omega}}$
2D	$C-\eta\left[ln\left\|\frac{\omega-\omega_0}{\omega_0}\right\|-1\right]\Theta(\omega-\omega_e)\Theta(\Omega_d)$	$C-\eta\left[Li_2\left(\frac{\Omega_d-\omega_0}{\omega-\omega_0}\right)-Li_2\left(\frac{\omega_0-\omega_e}{\omega_0-\omega}\right)-ln\,\frac{\omega_0-\omega_e}{\Omega_d-\omega_0}\,ln\,\frac{\omega_e-\omega}{\omega_0-\omega}\right]$
3D	$\chi\sqrt{\frac{\omega-\omega_e}{\Omega_C}}exp\left(-\frac{\omega-\omega_e}{\Omega_C}\right)\Theta(\omega-\omega_e)$	$\chi\left[\pi\sqrt{\frac{\omega_e-\omega}{\Omega_C}}exp\left(-\frac{\omega-\omega_e}{\Omega_C}\right)erfc\sqrt{\frac{\omega_e-\omega}{\Omega_C}}\right]$

The cognitive power here is the functional variable of the sphericity of a tree that is linked among the 1D, 2D, and 3D three-dimensional mode of its intelligence

This result suggested that the cognitive powers are indeed doable into the dynamic states of a plant's quantum field $\mathcal{P}_n^{(n_0)}(t)$ of $|n_0\rangle$ where the cognitive power generation rate $\mathcal{P}_n^{(n_0)}(t)$ in the prime state $|n_0 = 5\rangle$ and steady-state time limit of $\mathcal{P}_n^{(n_0)}(t \rightarrow \infty)$ transform it into physical energy.

Simulation of Physical Energy ($\mathbf{P_e}$) *of a Tree*

To convert cognitive power into physical energy of a tree, the quantum electrodynamics (*QeD*) of the plant has been implemented by initiating the local symmetry of the quantum field of a tree [14, 32, 33]. Hence, the gauge-field symmetries of the plant are being induced considering the momentum of the scalar field of the plant's quantum electrodynamics (*QeD*) field to transform cognitive power into physical energy by expressing Eq. 9.51:

$$E = F_{\mu\nu}F^{\mu\upsilon} + D_\mu\Phi^* \, D^\mu\Phi - V(\Phi^*\Phi), \tag{9.51}$$

where

$$\begin{aligned} D_\mu\Phi(x) &= \mathrm{iqA_\mu(x)\Phi\ (x)} \\ D_\mu\Phi^*(x) &= \partial_\mu\Phi^*(\mathrm{x}) - \mathrm{iqA_\mu(x)\Phi^*(x)}\text{'} \end{aligned} \tag{9.52}$$

and

$$V(\Phi^*\Phi) = \frac{\lambda}{2}\,(\Phi^*\Phi)^2 + m^2(\Phi^*\Phi). \tag{9.53}$$

Here, $\lambda > 0$ but $m^2 < 0$ therefore, $\Phi = 1$ is a local functional factor of the scalar field of tree, and thus, its quantum dynamics is $\Phi = \frac{v}{\sqrt{2}} * e^{i\theta}$ that can be expressed as

$$v = \sqrt{\frac{-\,2m^2}{\lambda}} \tag{9.54}$$

Here, the local U (1) symmetries into the electromagnetic field form the physical energy considering the scalar quantum dynamics of (x). Therefore, the quantum dynamics of (x) in terms of an activated scalar field of the physical energy of a tree in these local U (1) symmetries are being expressed as

$$\Phi\ (\mathrm{x}) = \frac{1}{\sqrt{2}}\ \Phi_\mathrm{r}(\mathrm{x}) * \mathrm{e}^{\mathrm{i}\Theta(\mathrm{x})}, \text{real } \Phi_\mathrm{r}(\mathrm{x}) > 0, \text{real } \Phi(\mathrm{x}). \tag{9.55}$$

Simplifying it by considering the momentum field (x) and the scalar field associated with the radial field of the tree, the physical energy can be rewritten as

$$V(\phi) = \frac{\lambda}{8}\left(\phi_r^2 - v^2\right)^2 \tag{9.56}$$

Subsequently, the momentum field of this physical energy can be shifted by applying the variable scalars, $\Phi_r(x) = v + \sigma(x)$, and can be expressed as

$$\phi_r^2 - v^2 = (v+\sigma)^2 - v^2 = 2v\sigma + \sigma^2 \tag{9.57}$$

$$V = \frac{\lambda}{8}\left(2v\sigma - \sigma^2\right)^2 = \frac{\lambda v^2}{2} * \sigma^2 + \frac{\lambda v}{2} * \sigma^3 + \frac{\lambda}{8} * \sigma^4. \tag{9.58}$$

Thus, the final functional derivative the physical energy $D_\mu\phi$ of a tree can be addressed as follows:

$$D_\mu\phi = \frac{1}{\sqrt{2}}\left(\partial_\mu\left(\phi_r e^{i\Theta}\right) + iqA_\mu * \phi_r e^{i\Theta}\right) = \frac{e^{i\Theta}}{\sqrt{2}}\left(\partial_\mu\phi_r + \phi_r * i\partial_\mu\Theta + \phi_r * iqA_\mu\right) \tag{9.59}$$

$$= \frac{1}{2}\left(\partial_\mu\,\sigma\right)^2 + \frac{(v+\sigma)^2}{2} * \left(\partial_\mu\Theta + qA_\mu\right)^2 \tag{9.60}$$

Simply, the calculative result of this physical energy suggested that plant's quantum electrodynamics (*QeD*) initiate the quantum field of a tree to transform it into the environmental energy considering plant's scalar field and its steady-state limit of $\partial_\mu\Theta + qA_\mu$.

Computation of Environmental Energy ($\mathbf{E_e}$)

Finally, to convert physical energy into environmental energy, the Higgs boson electromagnetic quantum dynamics is being used, considering local symmetry of U (1) analysis to use this energy is the primary source of the energy for everything [23, 29, 30]. Hence, this energy is derived from variable Higgs boson electromagnetic gauge particles; thus, considering the transformational laws of the scalar, it has been expressed as.

$$\begin{aligned} E_e \to D_\mu = \partial_\mu = ieA_\mu \quad & [\text{covariant derivatives}] \\ E_e = A_\mu + \frac{1}{e}\,\partial_\mu\alpha \quad & [A_\mu \text{ derivatives}] \end{aligned} \tag{9.61}$$

Here, the local U(1) gauge invariant is the environmental energy; thus, it has been considered the perplex scalar field particle and rewritten as follows:

$$E_e = (D^\mu)^\dagger \left(D_\mu \varnothing\right) - \frac{1}{4} F_{\mu\nu} F^{\mu\nu} - V(\varnothing). \tag{9.62}$$

As the environmental energy here is a quantum field of scalar particles, the clarification of this quantum field need to express by the kinetic mode of derivative the environmental energy which is

$$E_e(\eta, \xi) = (D^\mu \phi)^\dagger (D^\mu \phi) = (\partial^\mu + ieA^\mu)\phi^* \left(\partial_\mu - ieA_\mu\right) \phi \tag{9.63}$$

Here, $V(\eta, \xi) = \lambda\, v^2 \eta^2$ is the final term to the second order, and thus, the full environmental energy can be expressed as

$$E_e(\eta, \xi) = \frac{1}{2} \left(\partial_\mu \eta\right)^2 - \lambda v^2 \eta^2 + \frac{1}{2} \left(\partial_\mu \xi\right)^2 - \frac{1}{4} F_{\mu\nu} F^{\mu\nu} + \frac{1}{2} e^2 v^2 A_\mu^2 - evA_\mu (\partial^\mu \xi) \tag{9.64}$$

As the quantum field environmental energy subject to spontaneous change, the equation can be rewritten as a Lagrangian scalar to verify the creation of the environmental energy particles from a plant's quantum dynamics:

$$\begin{aligned} \mathcal{L}_{\text{scalar}} &= (D^\mu \phi)^\dagger (D^\mu \phi) - V\left(\phi^\dagger \phi\right) \\ &= (\partial^\mu + ieA^\mu) \frac{1}{\sqrt{2}} (v + h) \left(\partial_\mu - ieA_\mu\right) \frac{1}{\sqrt{2}} (v + h) - V\left(\phi^\dagger \phi\right) \end{aligned} \tag{9.65}$$

$$= \frac{1}{2} \left(\partial_\mu h\right)^2 + \frac{1}{2} e^2 A_\mu^2 (v + h)^2 - \lambda v^2 h^2 - \lambda v h^3 - \frac{1}{4} \lambda h^4 + \frac{1}{4} \lambda h^4 \tag{9.66}$$

Here, the Lagrangian scalar showed that the Higgs boson quantum dynamics can be surely started from the plant's environmental energy [18, 24, 26]. To verify and ascertain the plant's environmental energy generation, an additional computation taking into account its dispersed kinetic energy in relation to the angle θ from the plant's axis and the differential density and expressed as.

$$dn = \frac{1}{2} n(\in) \sin\theta d \in d\theta. \tag{9.67}$$

Rewriting these variables as integral over n and θ, it has been determined as

$$\frac{d\tau_{abs}}{dx} = \pi r_0^2 \left(\frac{m^2 c^4}{E}\right)^2 \int_{\frac{m^2 c^4}{E}}^{\infty} \overline{\in^{-2} n(\in)\, \phi[s_0(\in)]\, de,} \tag{9.68}$$

where

$$\overline{\phi}[s_0(\in)] = \int_1^{s_0(\in)} s\overline{\sigma}\,(s)ds, \overline{\sigma}(s) = \frac{2\sigma(s)}{\pi r_0^2}. \tag{9.69}$$

Simply, this calculation confirmed that the environmental energy generation from a plant can be represented by the equation [s0] to ascertain the net environmental energy formation as

$$\overline{\phi}[s_0] = 2s_0(\ln 4s_0 - 2) + \ln 4s_0(\ln 4s_0 - 2) - \frac{(\pi^2 - 9)}{3} + s_0^{-1}\left(\ln 4s_0 + \frac{9}{8}\right) + \ldots(s_0 \gg 1); \tag{9.70}$$

$$\overline{\phi}[s_0] = \left(\frac{2}{3}\right)(S_0 - 1)^{\frac{3}{2}} + \left(\frac{5}{3}\right)(S_0 - 1)^{\frac{5}{2}} - \left(\frac{1507}{420}\right)(S_0 - 1)^{\frac{7}{2}} + \ldots(s_0 - 1 \ll 1). \tag{9.71}$$

The function $\frac{\overline{\phi}[s_0]}{(s_0 - 1)}$ is revealed as $1 < s_0 < 10$; at larger s_0, it confirms a standard algorithm function of s_0. Thus, the energy spectra of the environmental energy can be written as.

$$n(\in) = 0, \in < \in_0 = C_{\in}^{-\alpha} \text{ or } D_{\in}^{\beta}, \in_0 < \in < \in_m = 0, \in > \in_m \tag{9.72}$$

Then, it can be transformed as

$$\left(\frac{d\tau_{abs}}{dx}\right)_\alpha = \pi r_0^2 C\left(\frac{m^2c^4}{E}\right)^{1-\alpha} \times \begin{cases} 0, & E < E_m \\ [F_\alpha(1) - F_\alpha(\sigma_m)], & E_m < E < E_0 \\ [F_\alpha(\sigma_0) - F_\alpha(\sigma_m)], & E > E_0; \end{cases} \tag{9.73}$$

$$\left(\frac{d\tau_{abs}}{dx}\right)_\beta = \pi r_0^2 D\left(\frac{m^2c^4}{E}\right)^{1+\beta} \times \begin{cases} 0, & E < E_m \\ [F_\beta(\sigma_m)], & E_m < E < E_0 \\ [F_\beta(\sigma_m) - F_\beta(\sigma_0)], & E > E_0. \end{cases} \tag{9.74}$$

Hence, the asymptotic formula correctly defines the plant's environmental energy spectrum; thus, variable energy generation in relation to quantum dynamics can be defined by the term as

$$\Gamma_\gamma \equiv \frac{dn_\gamma}{dVdt}. \tag{9.75}$$

This rate of environmental energy generation is being considered the variable energy, so, the Eq. 9.76, which expresses the environmental energy generation from a tree as

$$\frac{d\Gamma_{\gamma}^{LPM}}{d^3k} = \frac{d_F q_s^2 \alpha_{EM}}{4\pi^2 k} \int_{-\infty}^{\infty} \frac{dp_{\parallel}}{2\pi} \int \frac{d^2 \mathrm{p}_{\perp}}{(2\pi)^2} A\left(p_{\parallel}, k\right) \; \mathrm{Re}\left\{ 2\mathrm{P}_{\perp} \cdot f\left(\mathrm{p}_{\perp}; p_{\parallel}, k\right) \right\} \tag{9.76}$$

where dF is the functional state of the environmental energy (Nc in SU(Nc)), qs is the charge of the environmental energy (k ≡ |k|), and the kinetic functional mode is A(p||, k), and thus, the equation is being rewritten as

$$A\left(p_{\parallel}, k\right) \equiv \begin{cases} \dfrac{n_b\left(k + p_{\parallel}\right)\left[1 + n_b\left(p_{\parallel}\right)\right]}{2p_{\parallel}\left(p_{\parallel} + k\right)} & \text{scalars} \\[2ex] \dfrac{n_f\left(k + p_{\parallel}\right)\left[1 - n_f\left(p_{\parallel}\right)\right]}{2\left[p_{\parallel}\left(p_{\parallel} + k\right)\right]^2} \left[{p_{\parallel}}^2 + \left(p_{\parallel} + k\right)^2\right], & \text{fermions} \end{cases} \tag{9.77}$$

with

$$\begin{aligned} n_b(p) &\equiv \frac{1}{\exp[\beta(p - \mu)] - 1}, \\ n_f(p) &\equiv \frac{1}{\exp[\beta(p - \mu)] + 1} \end{aligned} \tag{9.78}$$

Then, the function $f(p\perp;\, p\|,\, k)$ is integrated to resolve the below equation to confirm the environmental energy generation from the tree by expressing following derivative as

$$\begin{aligned} 2\mathrm{p}_{\perp} = {} & i\delta E \mathrm{f}\left(\mathrm{p}_{\perp}; p_{\parallel}, k\right) + \frac{\pi}{2} C_F g_s^2 m_D^2 \int \frac{d^2 q_{\perp}}{(2\pi)^2} \frac{dq_{\parallel}}{2\pi} \frac{dq^0}{2\pi} 2\pi\delta\left(q^0 - q_{\parallel}\right) \\ & \times \frac{\mathrm{T}}{|\mathrm{q}|} \left[\frac{2}{|\mathrm{q}^2 - \Pi_{\mathrm{L}}(\mathrm{Q})|^2} + \frac{\left[1 - \left(q^0/|\mathrm{q}_{\parallel}|\right)^2\right]^2}{\left|(q^0)^2 - \mathrm{q}^2 - \Pi_{\mathrm{T}}(\mathrm{Q})\right|^2} \right] \left[f\left(\mathrm{p}_{\perp}; p_{\parallel}, k\right) - f\left(q + \mathrm{p}_{\perp}; p_{\parallel}, k\right) \right] \end{aligned} \tag{9.79}$$

Simply, this environmental energy generation is the derived from the explicit forms of environmental energy corresponding to its given the energy function of $f(p\perp;\, p\|,\, k)$ from a tree which can be ultimate point to mitigate any kind of energy needed for the planet.

Conclusions

The conduction of mathematical modeling of the intelligence of a tree is being simulated to confirm that plant has the enough cognitive power (C_p) to transform it into physical energy (P_e) and then environmental energy (E_e) which could be primacy sources of all energy for the planet. Hence, the computational simulation suggested that the attainability of cognitive power (C_p) is very much feasible in a tree which can be transfigured into physical energy by employing quantum electrodynamics (QeD) of the plant. Subsequently, this physical energy can further be transformed into environmental energy by deploying Higgs boson [BR ($H \rightarrow \gamma\gamma^{-}$)] electromagnetic fields of the plant. Simply, the transformation mechanism of cognitive power into physical energy and then environment energy shall indeed be the most groundbreaking scientific discovery in natural science to address global energy perplexity since it is abundant everywhere and benign to the ecosystem.

Acknowledgments The author declares that this research does not have any conflict of interest for publishing as a suitable book chapter.

References

1. Berges, J., & Mesterhazy, D. (2012). Introduction to the nonequilibrium functional renormalization group. *Nuclear Physics B-Proceedings Supplements, 228*, 37–60.
2. Bertozzi, E. (2010). Hunting the ghosts of a 'strictly quantum field': The Klein–Gordon equation. *European Journal of Physics, 31*(6), 1499.
3. Boettcher, I., Pawlowski, J. M., & Diehl, S. (2012). Ultracold atoms and the functional renormalization group. *Nuclear Physics B-Proceedings Supplements, 228*, 63–135.
4. Broz, M., Contreras, J. G., & Takaki, J. T. (2020). A generator of forward neutrons for ultra-peripheral collisions: nOOn. *Computer Physics Communications, 253*, 107181.
5. Chang, D. E., Sørensen, A. S., Demler, E. A., & Lukin, M. D. (2007). A single-photon transistor using nanoscale surface plasmons. *Nature Physics, 3*(11), 807–812.
6. Chen, G., Chen, S., Li, C., & Chen, Y. (2013). Examining non-locality and quantum coherent dynamics induced by a common reservoir. *Scientific Reports, 3*(1), 1–6.
7. Cheng, M., & Song, Y. (2012). Fano resonance analysis in a pair of semiconductor quantum dots coupling to a metal nanowire. *Optics Letters, 37*(5), 978–980.
8. Douglas, J. S., Habibian, H., Hung, C., Gorshkov, A. V., Kimble, H. J., & Chang, D. E. (2015). Quantum many-body models with cold atoms coupled to photonic crystals. *Nature Photonics, 9* (5), 326–331.
9. Dupuis, N., Canet, L., Eichhorn, A., Metzner, W., Pawlowski, J. M., Tissier, M., & Wschebor, N. (2021). The nonperturbative functional renormalization group and its applications. *Physics Reports, 910*, 1–114.
10. Eichler, J., & Stöhlker, T. (2007). Radiative electron capture in relativistic ion–atom collisions and the photoelectric effect in hydrogen-like high-Z systems. *Physics Reports, 439*(1–2), 1–99.
11. Englund, D., Majumdar, A., Faraon, A., Toishi, M., Stoltz, N., Petroff, P., & Vučković, J. (2010). Resonant excitation of a quantum dot strongly coupled to a photonic crystal nanocavity. *Physical Review Letters, 104*(7), 073904.

12. Fernández, J., & Martín, F. (2009). Electron and ion angular distributions in resonant dissociative photoionization of H2 and D2 using linearly polarized light. *New Journal of Physics, 11*(4), 043020.
13. Gazi, V., & Passino, K. M. (2005). Stability of a one-dimensional discrete-time asynchronous swarm. *IEEE Transactions on Systems, Man, and Cybernetics, Part B (Cybernetics), 35*(4), 834–841.
14. Hencken, K., Baur, G., & Trautmann, D. (2006). Transverse momentum distribution of vector mesons produced in ultraperipheral relativistic heavy ion collisions. *Physical Review Letters, 96*(1), 012303.
15. Hossain, M. F. (2016). Solar energy integration into advanced building design for meeting energy demand. *International Journal of Energy Research, 40*, 1293–1300.
16. Hossain, M. F. (2016). Theory of global cooling. *Energy, Sustainability, and Society, 6*, 24.
17. Hossain, M. F. (2017). Green science: Independent building technology to mitigate energy, environment, and climate change. *Renewable and Sustainable Energy Reviews, 73*, 695–705.
18. Hossain, M. F. (2018). Green science: Advanced building design technology to mitigate energy and environment. *Renewable and Sustainable Energy Reviews, 81*(2), 3051–3060.
19. Hossain, M. F. (2018). Photon energy amplification for the design of a micro-PV panel. *International Journal of Energy Research*. https://doi.org/10.1002/er.4118
20. Hossain, M. F. (2020). Modeling of global temperature control. *Environment, Development and Sustainability*. https://doi.org/10.1007/s10668-020-00924-6
21. Hossain, M. F. (2021). Sustainable building technology: Thermal control of solar energy to cool and heat the building naturally. *Environment, Development and Sustainability*. https://doi.org/10.1007/s10668-020-01212-z
22. Hossain, M. F. (2022). Ultraviolet germicidal irradiation (UVGI) application in building design to terminate pathogens naturally. *Material Today Sustainability*. https://doi.org/10.1016/j.mtsust.2022.100161
23. Jahnke, V., Luna, A., Patino, L., & Trancanelli, D. (2014). More on thermal probes of a strongly coupled anisotropic plasma. *Journal of High Energy Physics, 2014*(1), 1–40.
24. Jiaqi, L., Yi, Z., Chengkai, T., & Xingxing, Z. (2019). *INS aided high dynamic single-satellite position algorithm*. Paper presented at the 2019 IEEE international conference on signal processing, communications and computing (ICSPCC), pp. 1–5.
25. Johnson, B. R., Reed, M. D., Houck, A. A., Schuster, D. I., Bishop, L. S., Ginossar, E., et al. (2010). Quantum non-demolition detection of single microwave photons in a circuit. *Nature Physics, 6*(9), 663–667.
26. Naghiloo, M., Foroozani, N., Tan, D., Jadbabaie, A., & Murch, K. W. (2016). Mapping quantum state dynamics in spontaneous emission. *Nature Communications, 7*(1), 1–7.
27. Reinhard, P., & Nazarewicz, W. (2021). Nuclear charge densities in spherical and deformed nuclei: Toward precise calculations of charge radii. *Physical Review C, 103*(5), 054310.
28. Shalm, L. K., Hamel, D. R., Yan, Z., Simon, C., Resch, K. J., & Jennewein, T. (2013). Three-photon energy–time entanglement. *Nature Physics, 9*(1), 19–22.
29. Stehle, C., Zimmermann, C., & Slama, S. (2014). Cooperative coupling of ultracold atoms and surface plasmons. *Nature Physics, 10*(12), 937–942.
30. Szafron, R., & Czarnecki, A. (2016). High-energy electrons from the muon decay in orbit: Radiative corrections. *Physics Letters B, 753*, 61–64.
31. Tame, M. S., McEnery, K. R., Özdemir, Ş. K., Lee, J., Maier, S. A., & Kim, M. S. (2013). Quantum plasmonics. *Nature Physics, 9*(6), 329–340.
32. Ting, T. (2004). The polarization vector and secular equation for surface waves in an anisotropic elastic half-space. *International Journal of Solids and Structures, 41*(8), 2065–2083.
33. Tu, M. W., & Zhang, W. (2008). Non-markovian decoherence theory for a double-dot charge qubit. *Physical Review B, 78*(23), 235311.

34. Wang, C., Wang, W., & Chen, Z. (2017). *Single-satellite positioning algorithm based on direction-finding*. Paper presented at the 2017 Progress in electromagnetics research symposium-spring (PIERS), 2533–2538.
35. Yan, J. Z., & Chang, S. (2018). The contingent effects of political strategies on firm performance: A political network perspective. *Strategic Management Journal, 39*(8), 2152–2177.

Part IV
Sustainable Building

Chapter 10
Application of Solar Energy in Design Technology to Naturally Cool and Heat the Houses

Introduction

Traditional building heating and cooling methods have major detrimental effects on the atmosphere as it contributes to climate change crisis due to the burning of fossil fuels that releases billion of tons of CO_2 into the atmosphere [10, 29]. This, in turn, increases the devastating natural disasters and makes the earth more vulnerable [5, 16]. The ozone layer is also depleted due to the emissions of Chlorofluorocarbon (CFC) by conventional cooling mechanism of building globally [4, 13]. Hence, the ozone layer is a protective barrier that blocks high-frequency ultraviolet (UV) light from the sun [1, 20]. Unfortunately, the ozone layer is getting depleted, allowing ultraviolet (UV) radiation to reach the earth easily, and causing skin cancer for the mankind [18, 32]. Therefore, in this research, a sustainable advanced cooling and heating mechanism by employing helium-assisted photonic band gap and the Higgs boson (***H-B***) quantum activation by utilizing the curtain wall has been performed to decode the photons into cooling from photons and then heating-form photons, respectively. Hence, the photon emission inside the curtain wall forms a photonic band gap at *nano* point due to induced helium in the curtain wall; thus, the photonic quantum will be cooled to cool the inside premise naturally [2, 23]. Then, this cooling form photon can be converted into a heating-form photon by employing ***H–B*** ($H \rightarrow \gamma\gamma^{-}$) quantum to form an electromagnetic field to heat the premises naturally [7, 12, 28]. Simply, this natural mechanism of cooling and heating system shall be an outstanding scientific discovery to address the global energy, environment, and ozone layer perplexity significantly.

© The Author(s), under exclusive license to Springer Nature Switzerland AG 2025

M. F. Hossain, *Sustainable Global Development in Environment, Energy, Building, and Infrastructure*, https://doi.org/10.1007/978-3-031-84429-4_10

Methods and Simulation

Cooling Mechanism

The helium-assisted curtain wall is being used to decode the activated photon into cooling state one [9, 14, 21]. This helium-assisted curtain wall would establish a method of the integration of photonic band gap (PBG) defect arrays under nano point [13, 15, 26]. Hence, this nano point is considered photon reservoirs, thus, it has been analyzed by using MATLAB software to determine the precise formulation of the cooling form of photon through the helium-embedded curtain wall and expressed by the following Hamiltonian [11, 19, 34]:

$$H = \sum \omega_{ci} a_i^{\dagger} a_i + \sum_{K} \omega_k b_k^{\dagger} b_k + \sum_{ik} \left(V_{ik} a^{\dagger} b_k + V_{ik}^{*} b_k^{\dagger} a_i \right) \tag{10.1}$$

Here, the photon nano point modes are driven by the amplitude of the photonic point break among the photonic wavelength to cool the photos by employing two diodes and two resistors series in the curtain wall [24, 42, 43]. Thus, the current–voltage relations are being written as

$$I = I_L - I_O \left\{ \exp \left[\frac{q(V + I_{RS})}{AkTc} \right] - 1 \right\} - \frac{(V + I_{RS})}{R_{Sh}} \tag{10.2}$$

Here, I_L stands for the current at which the diode is saturated. *Rs* stands for series resistance, *A* stands for diode's function, *k* is Boltzmann's constant, *q* stands for the electrons, and T_C stands for the thermodynamic state of the cell. Therefore, the mathematical representation of the solar current represents the *I*–*q* relationship in photon cells of the curtain wall as follows:

$$\mathrm{I}_O = \mathrm{I}_{\mathrm{RS}} \left(\frac{T_C}{T_{\mathrm{ref}}} \right)^3 \exp \left[\frac{qEG\left(\frac{1}{T_{\mathrm{ref}}} - \frac{1}{T_C} \right)}{KA} \right] \tag{10.3}$$

Given the functional temperature and the speed of solar irradiation, let us indicate the solar current, and entering the photon cell is, therefore, the I–V relation for the photon module that is the sum of the I–V curves for each cell which can be rewritten as

$$V = IR_s + K_s \log \left[\frac{I_L - I + I_O}{I_O} \right] \tag{10.4}$$

Therefore,

$$V_{mo} = I_{mo}R_{Smo} + K_{mo}\log\left(\frac{I_{Lmo} - I_{mo} + I_{Omo}}{I_{Omo}}\right) \tag{10.5}$$

Here, I_{Lmo} stands for photonic current, I_{Omo} stands for activated current entering the diode, R_{smo} stands for series resistance, and K_{mo} stands for the factorial constant. Therefore, this equation can be modified considering the following module of I_{mo}–V_{mo} equation respecting the series connection of the cells.

As an analogy, once the NP cells are directly related in a series, configuration of the current–voltage relation can be restated as

$$V_{mo} = -I_{mo}\frac{Rs}{Np} + K\log\left(\frac{N_{sh}I_L - I_{mo} + N_pI_o}{N_pI_o}\right) \tag{10.6}$$

Eventually, this equation can be used to compute the current generation by photons considering the relativistic temperature conditions of the curtain wall cell:

$$I_L = G[I_{Sc} + K_I(T_{cool})] * V_{mo} \tag{10.7}$$

$$T_{cool} = \left(\frac{I_L}{(G * V_{mo}) \times (I_{sc} + K1)}\right)$$

Here, I_{sc} denotes the photonic current per unit area, K_I denotes the induced curtain wall cell, T_{cool} denotes the curtain wall cell's temperature, and G denotes the solar irradiance reaching the curtain wall in per unit area.

Heating Mechanism

The precision of a photon-heating equilibrium has been calculated by a thorough mathematical calculation by employing the Higgs–Boson electromagnetic field to transform cooling state photon into a heating state photon inside the curtain wall skin [30, 36, 41]. Here, the Higgs–Boson electromagnetic field will form a local abelian symmetry rapidly, therefore, the symmetry of the gauge field will be transformed into a complex scalar field of heating particle as

$$\mathcal{L} = -\frac{1}{4}F_{\mu\nu}F^{\mu\nu} + D_\mu\Phi^* D^\mu\Phi - V(\Phi^*\Phi) \tag{10.8}$$

where

$$D_{\mu}\Phi(x) = \partial_{\mu}\Phi(\mathrm{x0}) + \mathrm{iqA_{\mu}}(\mathrm{x0})\Phi\ (\mathrm{x0})$$
$$D_{\mu}\Phi^{*}(x) = \partial_{\mu}\Phi^{*}(\mathrm{x0}) - \mathrm{iqA_{\mu}}(\mathrm{x0})\Phi^{*}(\mathrm{x0}) \tag{10.9}$$

and

$$V(\Phi^{*}\Phi) = \frac{\lambda}{2}\ (\Phi^{*}\Phi)^{2} + m^{2}(\Phi^{*}\Phi) \tag{10.10}$$

Thus, the scalar field's generation and the complex phase (x) would be the source for the creation of a scalar. Thus, the phase of (x) in the context of local U (1) symmetry denotes both the expectation value's phase and the dependent-on x dynamical (x) field's phase.

To confirm this statistic, a polar coordinate has been employed in those phases and thus, the equation is being rewritten as

$$\Phi\ (\mathrm{x}) = \frac{1}{\sqrt{2}}\ \Phi_{\mathrm{r}}(\mathrm{x}) * \mathrm{e}^{\mathrm{i}\Theta(\mathrm{x})}, \text{real } \Phi_{\mathrm{r}}(\mathrm{x}) > 0, \text{real } \Phi(\mathrm{x}) \tag{10.11}$$

Once the expectant value comes to the real fields, the (x), the field can be determined as the vectorial potential as follows:

$$V(\phi) = \frac{\lambda}{8}\ \left(\phi_{r}^{2} - v^{2}\right)^{2} + \text{const}, \tag{10.12}$$

or transformed by its vector, $\Phi_{\mathrm{r}}(\mathrm{x}) = \mathrm{v} + \sigma(\mathrm{x})$,

$$\phi_{r}^{2} - v^{2} = (v + \sigma)^{2} - v^{2} = 2v\sigma + \sigma^{2} \tag{10.13}$$

$$V = \frac{\lambda}{8}\ \left(2v\sigma - \sigma^{2}\right)^{2} = \frac{\lambda v^{2}}{2} * \sigma^{2} + \frac{\lambda v}{2} * \sigma^{3} + \frac{\lambda}{8} * \sigma^{4} \tag{10.14}$$

Here, the covariant derivative $D_{\mu}\phi$ becomes

$$D_{\mu}\phi = \frac{1}{\sqrt{2}}\ \left(\partial_{\mu}\left(\phi_{r}e^{i\Theta}\right) + iqA_{\mu} * \phi_{r}e^{i\Theta}\ \right) = \frac{e^{i\Theta}}{\sqrt{2}}\ \left(\partial_{\mu}\phi_{r} + \phi_{r} * i\partial_{\mu}\Theta + \phi_{r} * iqA_{\mu}\right) \tag{10.15}$$

$$\begin{aligned}\left|D_{\mu}\phi\right|^{2} &= \frac{1}{2}\ \left|\partial_{\mu}\phi_{r} + \phi_{r} * i\partial_{\mu}\Theta + \phi_{r} * iqA_{\mu}\right|^{2} = \frac{1}{2}\ \left(\partial_{\mu}\phi_{r}\right) + \frac{\phi_{r}^{2}}{2} * \left(\partial_{\mu}\Theta qA_{\mu}\right)^{2} \\ &= \frac{1}{2}\ \left(\partial_{\mu}\ \sigma\right)^{2} + \frac{(v + \sigma)^{2}}{2} * \left(\partial_{\mu}\Theta + qA_{\mu}\right)^{2}\end{aligned} \tag{10.16}$$

Altogether,

$$\mathcal{L}_{\text{heat}} = \frac{1}{2}\left(\partial_\mu \sigma\right)^2 - \frac{\lambda v^2}{2} * \sigma^2 - \frac{1}{4} F_{\mu\nu} F^{\mu\nu} + \frac{v^2}{2} * \left(qA_\mu + \partial_\mu \Theta\right)^2 \tag{10.17}$$

Here, this heating ($\mathcal{L}_{\text{heat}}$) energy formation due to the Higgs–Boson electromagnetic field suggests that scalar particle λv^2 has been evolved within that initiate to releasing heating energy inside the curtain wall to heat the premises.

Results and Discussion

Cooling Mechanism

To confirm the cooling photon generation by the helium-induced curtain wall, a quantitative mathematical analysis has been conducted by using MATLAB software [25, 39, 40]. As the curtain wall has the energy-coupling limit, the cooling photon proliferation production has been calculated by the clarification of all dynamic photon respecting the modes of 1D, 2D, and 3D (*see* Table 10.1).

Hence, the helium-induced curtain wall helped to form the PBG into the curtain wall, thus, the density of states (DOS) the cooling photovoltaic energy is being calculated within its quantum area as:

$$J(\omega) = \varrho(\omega)|V(\omega)|^2 \tag{10.18}$$

Using the photon dynamics in Eq. 10.18, the cooling photon frequency and proliferative photon dynamics have been quantified by using the following equation:

$$u(t, t_0) = \frac{1}{1 - \Sigma'(\omega_b)} e^{-i\omega(t - t_0)} + \int_{\omega_e}^{\infty} d\omega \frac{J(\omega) e^{-i\omega(t - t_0)}}{[\omega - \omega_c - \Delta(\omega)]^2 + \pi^2 J^2(\omega)} \tag{10.19}$$

where $\Sigma'(\omega_b) = [\partial\Sigma(\omega)/\partial\omega]_{\omega = \omega_b}$ and $\Sigma(\omega)$ denote the reservoir-induced PB array,

$$\Sigma(\omega) = \int_{\omega_e}^{\infty} d\omega' \frac{J(\omega')}{\omega - \omega'} \tag{10.20}$$

Here, the photonic bandgap of the cooling photonic frequency mode is being calculated using the pole condition of photon reservoir, which is an integral function of the cooling photon particles.

As the cooling photons are being generated during the transition from the PBG to the crossover region of the curtain wall, the photonic band gap is being linked to the criterion of the photonic band edge [3, 6, 27]. Then, cooling state photon energy

Table 10.1 The photonic cooling dynamics into the curtain wall in different dimensional modes are being clarified considering the dynamics of photons of 1D, 2D, and 3D entering the curtain wall linked to its distinct unit area

Photon	Unit area $J(\omega)$ for different DOS	Reservoir-induced self-energy correction $\Sigma(\omega)$
1D	$\frac{C}{\pi}\frac{1}{\sqrt{\omega-\omega_e}}\Theta(\omega-\omega_e)$	$-\frac{C}{\sqrt{\omega_e-\omega}}$
2D	$-\eta\left[ln\left\|\frac{\omega-\omega_0}{\omega_0}\right\|-1\right]\Theta(\omega-\omega_e)\Theta(\Omega_d-\omega)$	$\eta\left[Li_2\left(\frac{\Omega_d-\omega_0}{\omega-\omega_0}\right)-Li_2\left(\frac{\omega_0-\omega_e}{\omega_0-\omega}\right)-ln\,\frac{\omega_0-\omega_e}{\Omega_d-\omega_0}\,ln\,\frac{\omega_e-\omega}{\omega_0-\omega}\right]$
3D	$\chi\sqrt{\frac{\omega-\omega_e}{\Omega_C}}exp\left(-\frac{\omega-\omega_e}{\Omega_C}\right)\Theta(\omega-\omega_e)$	$\chi\left[\pi\sqrt{\frac{\omega_e-\omega}{\Omega_C}}exp\left(-\frac{\omega-\omega_e}{\Omega_C}\right)erfc\sqrt{\frac{\omega_e-\omega}{\Omega_C}}-\sqrt{\pi}\right]$

determination, which is theoretically derived from real-time quantum feedback control, has been explicitly calculated by the following equation:

$$\rho(t) = \sum_{n=0}^{\infty} \mathcal{P}_n^{(n_0)}(t)|n_0\rangle\langle n_0| \tag{10.21}$$

$$\mathcal{P}_n^{(n_0)}(t) = \frac{[v(t,t)]^n}{[1+v(t,t)]^{n+1}}[1-\Omega(\mathrm{t})]^{n_0} \times \sum_{k=0}^{\min\{n_0,n\}} \binom{n_0}{k}\binom{n}{k}\left[\frac{1}{v(t,t)}\frac{\Omega(\mathrm{t})}{1-\Omega(\mathrm{t})}\right]^k \tag{10.22}$$

Here, $\Omega(\mathrm{t}) = \frac{|u(t,t_0)|^2}{1+v(t,t)}$ is related to time-induced factor, which is the Fock state cooling photon, is being induced into dynamic states of $|n_0\rangle is\ \mathcal{P}_n^{(n_0)}(t)$ cooling photon. Hence, the formation of photon $\mathcal{P}_n^{(n_0)}(t)$ in the functional state $|n_0 = t\rangle$ has the steady-state limit, $\mathcal{P}_n^{(n_0)}(t \to \infty\ )$, which is the dynamics of the generation of cooling photon energy to cool the premises inside the curtain wall.

Heating Mechanism

As the induced H–B [BR (H $\rightarrow \gamma\gamma^-$] quantum field of the curtain wall creates the electromagnetic field, the quantum gauge variant would force the gauge particle $\varnothing' \rightarrow e^{i\alpha(x)}\varnothing$ to convert cooling-form photons into heating-form photons that can be explained by implementing covariant derivative of the following scalar field [8, 31, 37]:

$$\begin{aligned} \partial_\mu \rightarrow D_\mu = \partial_\mu = ieA_\mu \quad & [\text{covariant derivatives}] \\ A'_\mu = A_\mu + \frac{1}{e}\ \partial_\mu\alpha \quad & [A_\mu \text{ derivatives}] \end{aligned} \tag{10.23}$$

Here, the local U(1) variant which is a *Lagrangian* of a scalar field:

$$\mathcal{L} = (D^\mu)^\dagger\ (D_\mu\varnothing) - \frac{1}{4}\ F_{\mu\nu}F^{\mu\nu} - V(\varnothing) \tag{10.24}$$

Addition to the Lagrangian, this term stands for the kinetic part of the gauge field of heating photon array which is the quantum field of impacted Lagrangian, which is the kinetic term that gives rise to massive scalar particles as:

$$\text{Kinetic term}: \mathcal{L}_{kin}(\eta,\xi) = (D^\mu\phi)^\dagger(D^\mu\phi) = (\partial^\mu + ieA^\mu)\phi^*\ (\partial_\mu - ieA_\mu)\ \phi \tag{10.25}$$

and

$$\mathcal{L}_{\text{kin}}(\eta,\xi)=\frac{1}{2}\left(\partial_{\mu}\eta\right)^{2}-\lambda v^{2}\eta^{2}+\frac{1}{2}\left(\partial_{\mu}\xi\right)^{2}-\frac{1}{4}F_{\mu\nu}F^{\mu\nu}+\frac{1}{2}e^{2}v^{2}A_{\mu}^{2}-evA_{\mu}\left(\partial^{\mu}\xi\right)+\text{int. terms} \tag{10.26}$$

Here, the particles of the heating photon, in this case, are massive, but they also incorporated by a quantum mass that enable the spectra of heated photon particles within the quantum field as

$$\mathcal{L}_{\text{scalar}}=(D^{\mu}\phi)-V\left(\phi^{\dagger}\phi\right)=\left(\partial^{\mu}+ieA^{\mu}\right)\frac{1}{\sqrt{2}}(v+h)\left(\partial_{\mu}-ieA_{\mu}\right)\frac{1}{\sqrt{2}}(v+h)-V\left(\phi^{\dagger}\phi\right) \tag{10.27}$$

$$=\frac{1}{2}\left(\partial_{\mu}h\right)^{2}+\frac{1}{2}e^{2}A_{\mu}^{2}(v+h)^{2}-\lambda v^{2}h^{2}-\lambda vh^{3}-\frac{1}{4}\lambda h^{4}+\frac{1}{4}\lambda h^{4} \tag{10.28}$$

Here, the modification of the functions of the quantum field heating photons by integration of scalar in it can be calculated as.

$$\frac{d\tau_{abs}}{dx}=\pi r_{0}^{2}\left(\frac{m^{2}c^{4}}{E}\right)^{2}\int_{\frac{m^{2}c^{4}}{E}}^{\infty}\in^{-2}n(\in)\overline{\varphi}[s_{0}(\in)]de, \tag{10.29}$$

where

$$\overline{\varphi}[s_{0}(\in)]=\int_{1}^{s_{0}(\in)}s\overline{\sigma}(s)ds,\overline{\sigma}(s)=\frac{2\sigma(s)}{\pi r_{0}^{2}}. \tag{10.30}$$

Hence, the dimensional variable and dimensionless cross-section have been identified for the values between 1 and 10, which is the variable [s0] of heating photons, calculated using a whole functional asymptotic array of the curtain wall as.

$$\overline{\varphi}[s_{0}]=\frac{1+\beta_{0}^{2}}{1-\beta_{0}^{2}}\ln\omega_{0}-\beta_{0}^{2}\ln\omega_{0}-\ln^{2}\omega_{0}-\frac{4\beta_{0}}{1-\beta_{0}^{2}}+2\beta_{0}+4\ln\omega_{0}\ln(\omega_{0}+1)-L(\omega_{0}), \tag{10.31}$$

where.

$$\beta_{0}^{2}=\frac{1-1}{s_{0}},\ \omega_{0}=\frac{(1+\beta_{0})}{(1-\beta_{0})},$$
$$L(\omega_{0})=\int_{1}^{\omega_{0}}\omega^{-1}\ln(\omega+1)d\omega. \tag{10.32}$$

Here, their variable function can be written as.

$$(\omega+1)=\omega\left(\frac{1+1}{\omega}\right), L(\omega_0)=\frac{1}{2}\ln^2\omega_0+L'(\omega_0),$$

where

$$L'(\omega_0)=\int_1^{\omega_0}\omega^{-1}\ln\left(1+\frac{1}{\omega}\right)d\omega,$$
$$=\frac{\pi^2}{12}-\sum_{n=1}^{\infty}(-1)n^{-1}n^{-2}\omega_0^{-n}. \tag{10.33}$$

The analysis of heating photon spectra is, therefore, a functional area for the asymptotic gauge that derived from this [BR (H → γγ¯] quantum field which is the driving mechanism to generate heating photon per unit volume, as measured by bremsstrahlung processes as [22, 33, 35].

$$\Gamma_\gamma\equiv\frac{dn_\gamma}{dVdt} \tag{10.34}$$

To confirm the thermodynamically heating photon generation at temperature T, it can derived by the following equation:

$$\frac{d\Gamma_\gamma^{LPM}}{d^3k}=\frac{d_F q_s^2\alpha_{EM}}{4\pi^2k}\int_{-\infty}^{\infty}\frac{dp_\parallel}{2\pi}\int\frac{d^2\mathrm{p}_\perp}{(2\pi)^2}A\left(p_\parallel,k\right)\ \mathrm{Re}\left\{2\mathrm{P}_\perp\cdot\mathrm{f}\left(\mathrm{p}_\perp;p_\parallel,k\right)\right\} \tag{10.35}$$

Here, d_F denotes the functional quantum particle [N_c in SU(N_c)], q_s denotes the abelian gauge field, k ≡ |k| denotes the kinetic mode under the A(p||, k) of the emitted particle as:

$$A\left(p_\parallel,k\right)\equiv\begin{cases}\dfrac{n_b\left(k+p_\parallel\right)\left[1+n_b\left(p_\parallel\right)\right]}{2p_\parallel\left(p_\parallel+k\right)}, & \text{scalars}\\[2ex] \dfrac{n_f\left(k+p_\parallel\right)\left[1-n_f\left(p_\parallel\right)\right]}{2\left[p_\parallel\left(p_\parallel+k\right)\right]^2}\left[p_\parallel{}^2+\left(p_\parallel+k\right)^2\right], & \text{fermions}\end{cases} \tag{10.36}$$

with

$$n_b(p) \equiv \frac{1}{\exp[\beta(p-\mu)]-1}, \quad n_f(p) \equiv \frac{1}{\exp[\beta(p-\mu)]+1} \tag{10.37}$$

Hence, in the transformation of the heating photons into the heating state energy, the following linear integral equation was being conducted [17, 35, 38].

$$\begin{aligned} 2\mathrm{p}_\perp = & \, i\delta E\, \mathrm{f}\left(\mathrm{p}_\perp; p_\parallel, k\right) + \frac{\pi}{2} C_F g_s^2 m_D^2 \int \frac{d^2 q_\perp}{(2\pi)^2} \frac{dq_\parallel}{2\pi} \frac{dq^0}{2\pi} 2\pi\delta\left(q^0 - q_\parallel\right) \\ & \times \frac{\mathrm{T}}{|\mathrm{q}|} \left[\frac{2}{\left|\mathrm{q}^2 - \Pi_\mathrm{L}(\mathrm{Q})\right|^2} + \frac{\left[1 - \left(q^0/\left|\mathrm{q}_\parallel\right|\right)^2\right]^2}{\left|\left(q^0\right)^2 - \mathrm{q}^2 - \Pi_\mathrm{T}(\mathrm{Q})\right|^2} \right] \\ & \times \left[\mathrm{f}\left(\mathrm{p}_\perp; p_\parallel, k\right) - \mathrm{f}\left(\mathrm{q} + \mathrm{p}_\perp; p_\parallel, k\right)\right] \end{aligned} \tag{10.38}$$

Subsequently, the heating photon energy dynamics is determined by considering the state of thermodynamic of the curtain wall and the difference between the quasi-particle energies, δE, as computed by the following equations:

$$\delta E \equiv k^0 + E_\mathrm{p} \mathrm{sign}\left(p_\parallel\right) - E_{\mathrm{p+k}} \mathrm{sign}\left(p_\parallel + k\right) \tag{10.39}$$

Simplifying the equation by shifting the explicit form of δE into the definite form, it can be rewritten as

$$\delta E = \left[\frac{\mathrm{p}_\perp^2 + m_\infty^2}{2}\right] \left[\frac{k}{p_\parallel\left(k + p_\parallel\right)}\right] \tag{10.40}$$

Thus, the result here confirms a variable of heating photon energy generation considering the leading order heating photon release rate from the curtain wall considering the thermal photon energy spectral density to heat the promises inside the curtain wall time as a function of thermal photon (*see* Fig. 10.1).

Conclusions

The traditional heating and cooling systems employed in housing represent considerable worldwide challenges, as they contribute to climate change and threaten severe damage to the ozone layer. To mitigate these detrimental effects, a mathematical assessment has been performed employing computerized modeling to transform photon into the cooling state through the implementation of the helium-assisted curtain wall to cool the house naturally. Furthermore, the *H–B* [BR (H → γγ¯]

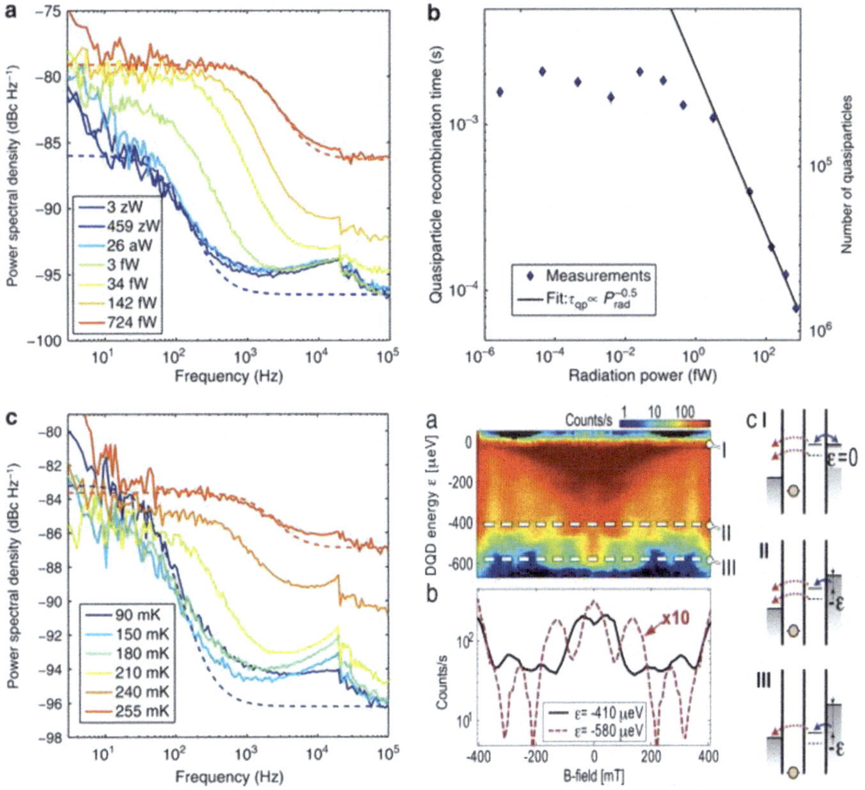

Fig. 10.1 (**a**) The frequency versus amplitude spectral density of thermal photon energy frequency. (**b**) The frequency in the recorded spectra that linked between the quasi-particle recombination time. (**c**) Examining the thermal photon power spectrum density at different frequencies. (**d**) The quantum B-field of the Higgs–Boson related to the average energy measurement of heating photons (DQD) to heat the premises inside the captain wall

quantum field has been employed to generate electromagnetic field in the curtain wall to convert cooling photon into heating photon to naturally heat the houses. Simply, this natural mechanism of cooling and heating will represent a novel scientific domain aimed at deciphering photon thermodynamics to convert photons into a cooling state *(HcP⁻)* and subsequently transform these cooling state photons into heating state photons *(HtP⁻)* to address the global energy, environmental and ozone layer crises significantly.

Acknowledgments The author declares that this chapter does have any conflict of interest for publishing by a quality publisher.

References

1. Agger, A. K., & Sørensen, A. H. (1997). Atomic and molecular structure and dynamics. *Physical Review A, 55*, 402.
2. Arnold, P. (2001). Photon emission from ultrarelativistic plasmas. *Journal of High Energy Physics*.
3. Artemyev, N., Jentschura, U. D., Serbo, V. G., & Surzhykov, A. (2012). Strong electromagnetic field EFFECTS in ultra-relativistic heavy-ion collisions. *European Physical Journal C, 72*, 1935.
4. Baur, G., Hencken, K., Trautmann, D., Sadovsky, S., & Kharlov, Y. (2002). Dense laser-driven electron sheets as relativistic mirrors for coherent production of brilliant X-ray and γ-ray beams. *Physics Reports, 364*, 359.
5. Baur, G., Hencken, K., & Trautmann, D. (2007). Revisiting unitarity corrections for electromagnetic processes in collisions of relativistic nuclei. *Physics Reports, 453*, 1.
6. Becker, U., Grün, N., & Scheid, W. (1987). K-shell ionisation in relativistic heavy-ion collisions. *Journal of Physics B: Atomic and Molecular Physics, 20*, 2075.
7. Belkacem, H., Gould, B., Feinberg, R., Bossingham, W. E., & Meyerhof. (1993). Semiclassical dynamics and relaxation. *Physical Review Letters, 71*, 1514.
8. Benavides, N. D., & Chapman, P. L. (2008). Modeling the effect of voltage ripple on the power output of photovoltaic modules. *IEEE Transactions on Industrial Electronics, 55*(7), 2638–2643.
9. Boukhezzar, B., & Siguerdidjane, H. (2009). Nonlinear control with wind estimation of a DFIG variable speed wind turbine for power capture optimization. *Energy Conversion and Management, 50*(4), 885–892.
10. Celik, A. N., & Acikgoz, N. (2007). Modelling and experimental verification of the operating current of mono-crystalline photovoltaic modules using four- and five-parameter models. *Applied Energy, 84*(1), 1–15.
11. De Soto, W., Klein, S. A., & Beckman, W. A. (2006). Improvement and validation of a model for photovoltaic array performance. *Solar Energy, 80*(1), 78–88.
12. Douglas, J. S., Habibian, H., Hung, C.-L., Gorshkov, A. V., Kimble, H. J., & Chang, D. E. (2015). Quantum many-body models with cold atoms coupled to photonic crystals. *Nature Photonics*.
13. Eichler, J., & Stöhlker, T. (2007). Radiative electron capture in relativistic ion-atom collisions and the photoelectric effect in hydrogen-like high-Z systems. *Physics Reports, 439*, 1.
14. Faida, H., & Saadi, J. (2010). Modelling, control strategy of DFIG in a wind energy system and feasibility study of a wind farm in Morocco. *International Review on Modelling and Simulations (IREMOS), 3*(6), 1350–1362.
15. Ghennam, T., Berkouk, E. M., & Francois, B. (2007). A vector hysteresis current control applied on three-level inverter. Application to the active and reactive power control of doubly fed induction generator based wind turbine. *International Review of Electrical Engineering (IREE), 2*(2), 250–259.
16. Gopal, C., Mohanraj, M., Chandramohan, P., & Chandrasekar, P. Renewable energy source water pumping systems-A literature review. *Renewable and Sustainable Energy Reviews*. https://doi.org/10.1016/j.rser.2013.04.012
17. Gould, R. J. (1967). Pair production in photon-photon collisions. *Physical Review*.
18. Gupta, N., Singh, S. P., Dubey, S. P., & Palwalia, D. K. (2011). Fuzzy logic controlled three-phase three-wired shunt active power filter for power quality improvement. *International Review of Electrical Engineering (IREE), 6*(3), 1118–1129.
19. Hencken, K. (2006). Transverse momentum distribution of vector mesons produced in Ultraperipheral relativistic heavy ion collisions. *Physical Review Letters*.
20. Hossain, F. (2016). Solar energy integration into advanced building design for meeting energy demand and environment problem. *Journal of Energy Research, 17*, 49–55.

21. Kamal, E., Koutb, M., Sobaih, A. A., & Abozalam, B. (2010). An intelligent maximum power extraction algorithm for hybrid wind-diesel-storage system. *International Journal of Electrical Power & Energy Systems, 32*(3), 170–177.
22. Klein, S. A. (1975). Calculation of flat-plate collector loss coefficients. *Solar Energy, 17*, 79–80.
23. Langer, L., et al. (2014). Access to long-term optical memories using photon echoes retrieved from semiconductor spins. *Nature Photonics, 8*, 851–857.
24. Li, Q., Xu, D. Z., Cai, C. Y., & Sun, C. P. (2013). Recoil effects of a motional scatterer on single-photon scattering in one dimension. *Scientific Reports*.
25. Najjari, B., Voitkiv, A. B., Artemyev, A., & Surzhykov, A. (2009). Simultaneous electron capture and bound-free pair production in relativistic collisions of heavy nuclei with atoms. *Physical Review A, 80*, 012701.
26. Park, J., Kim, H.-g., Cho, Y., & Shin, C. (2014). Simple modeling and simulation of photovoltaic panels using Matlab/Simulink. *Advanced Science and Technology Letters, 73*, 147–155.
27. Pregnolato, T., Lee, E. H., Song, J. D., Stobbe, S., & Lodahl, P. (2015). Single-photon nonlinear optics with a quantum dot in a waveguide. *Nature Communications*.
28. Reinhard, A. (2011). Strongly correlated photons on a chip. *Nature Photonics*.
29. Robyns, B., Francois, B., Degobert, P., & Hautier, J. P. (2012). *Vector control of induction machines*. Springer.
30. Şengül, M. Y., et al. (1999). Electromagnetic lepton pair production in relativistic heavy-ion collisions. *Annals of Physics, 272*, 7.
31. Sharma, K. G., Bhargava, A., & Gajrani, K. (2013). Stability analysis of DFIG based wind turbines connected to electric grid. *International Review on Modelling & Simulations*.
32. Sivasankar, G., & Kumar, V. S. (2013). Improving Low voltage ride through of wind generators using STATCOM under symmetric and asymmetric fault conditions. *International Review on Modelling & Simulations*.
33. Soedibyo, P., Agung, F., & Ashari, M. (2013). Grid quality hybrid power system control of microhydro, wind turbine and fuel cell using fuzzy logic. *International Review on Modelling & Simulations*.
34. Soon, J. J., & Low, K.-S. (2012). Optimizing photovoltaic model parameters for simulation. In IEEE international symposium on industrial electronics.
35. Tame, M. S., McEnery, K. R., Özdemir, Ş. K., Lee, J., Maier, S. A., & Kim, M. S. (2013). Quantum plasmonics. *Nature Physics*.
36. Tan, Y. T., Kirschen, D. S., & Jenkins, N. (2004). A model of PV generation suitable for stability analysis. *IEEE Transactions on Energy Conversion, 19*(4), 748–755.
37. Tu, M. W. Y., & Zhang, W. M. (2008). Non-Markovian decoherence theory for a double- dot charge qubit. *Physical Review B, 78*, 235311.
38. Valluri, S. R., Becker, U., Grün, N., & Scheid, W. (1984). Relativistic collisions of highly-charged ions. *Journal of Physics B: Atomic and Molecular Physics, 17*, 4359.
39. Xiao, W., Dunford, W. G., & Capal, A. (2004). A novel modeling method for photovoltaic cells. In *35th annula IEEE power electronics specialists conference*, Aachen, Germany, pp. 1950–1956.
40. Xiao, Y. F., et al. (2010). Asymmetric Fano resonance analysis in indirectly coupled microresonators. *Physical Review A, 82*, 065804.
41. Yan, W.-B., & Fan, H. (2014). Single-photon quantum router with multiple output ports. *Scientific Reports*.
42. Yang, L., Wang, S., Zeng, Q., Zhang, Z., Pei, T., Li, Y., & Peng, L.-M. (2011). Efficient photovoltage multiplication in carbon nanotubes. *Nature Photonics, 5*, 672–676.
43. Zhu, Y., Hu, X., Yang, H., & Gong, Q. (2014). On-chip plasmon-induced transparency based on plasmonic coupled nanocavities. *Scientific Reports*.

Chapter 11
Application of Printable Solar Cell in the Building Design to Mitigate Net Energy Demand Naturally

Introduction

Traditional energy utilization by the building sectors is believed to be responsible for 40% of the climate change crisis [2, 3, 47]. In conventional sources of energy at this moment, there is a total of 36,600 EJ of fossil fuels that have been reserved all over the world [1, 6]. For instance, crude oil is 1.65×10^{11} tons, natural gas is 1.81×10^{14} tons, high-quality coal is 4.90×10^{11} tons, and low-quality coal is 4.3×10^{11} tons [4, 7]. Over the course of the years 1980–1990, 2000–2004, 1994–1995, and 2025–2026, the annual levels of energy consumption across the globe varied from 283 EJ to 590 EJ [12, 17]. A total of 560 EJ of fossil energy was consumed in 2023, with the construction industry being responsible for 224 EJ of that amount and this fossil fuel will run out within the next 65 years. Considering the imminent environmental catastrophe that will be brought about by the combustion of fossil fuels, the utilization of sustainable and renewable energy sources is necessary for the continued existence of the planet. Interestingly, "solar radiation" refers to a renewable energy source would be the best option that may be trapped by incorporating printable solar cells installed on the exterior curtain wall of the building to address its required energy demand. Recent study shows that the average quantity of solar radiation that reaches the earth is 5,460,000 EJ which is nearly 10,000 times more than the total energy needed by the entire world. If only around 0.01% of solar energy that reaches the earth is used, it will satisfy the net energy requirements of the world. Therefore, in this study, printable solar panel is being proposed to install at the outer layer of curtain wall skin to trap the solar energy and convert them into clean energy, thereby satisfying the needs of the net energy demand for that building which is abundant and benign to the environment.

© The Author(s), under exclusive license to Springer Nature Switzerland AG 2025

M. F. Hossain, *Sustainable Global Development in Environment, Energy, Building, and Infrastructure*, https://doi.org/10.1007/978-3-031-84429-4_11

Methods and Materials

A sustainable technology for building design has been presented, wherein 25% of the external curtain wall would be adorned with printed solar cells to capture solar energy, fulfilling the building's net energy requirements. Necessarily, it is essential to determine the accurate angle, latitude, longitude, and the Cartesian coordinate system to optimize the printable solar cells installed on the exterior curtain wall skin to trap solar energy more efficiently [4, 5, 7]. The module should be oriented in the direction of the north for the southern hemisphere and the south for the northern hemisphere [10, 18, 21]. Hence, the horizon system needs to be articulated using Cartesian coordinates, where x denotes south, y signifies west, and z indicates the zenith and the azimuth angle, and the altitude needs to be in the two angles to determine the physical positions of the celestial body [8, 11, 13]. Naturally, the Cartesian coordinates of the equatorial system for the printable solar cells need to be precise considering that the z-axis is aligned with north, the y-axis is parallel to the east–west axis, and the x-axis is perpendicular to the other two axes to ascertain the position of the celestial object accurately (*see* Fig. 11.1).

Following the determination of the angles and Cartesian coordinates of the printable solar cells (*PSC*), the wiring system is being installed to connect with the printable solar cells to convert direct current (DC) photovoltaic current into alternating current (AC) from the solar irradiance [9, 20, 34]. Hence, the analysis of the single-diode equivalent circuit mode of the printable solar cells has been conducted for measuring the current–voltage (I–V) characteristic parameters [32, 35, 43]. Following this, a minimum control panel is being utilized to test the photovoltaic energy precisely considering the parameters of DC/AC converter, resistance curves, current load, and charge controller (*see* Fig. 11.2).

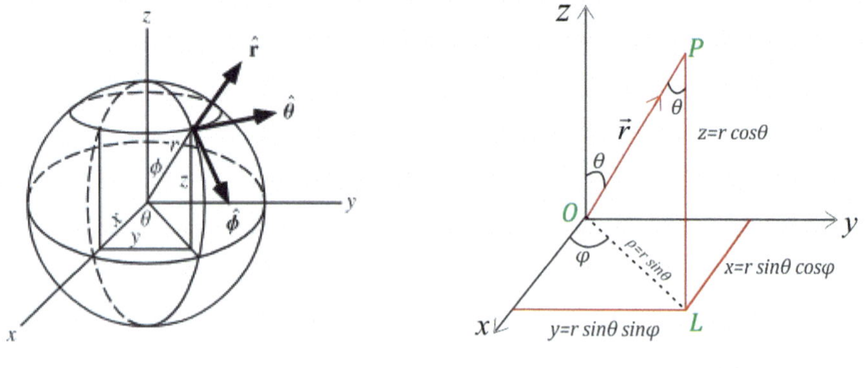

(a) (b)

Fig. 11.1 (**a**) The angle of a printed solar cell has been analyzed, considering the longitude and latitude angle, latitude, longitude, and (**b**) the Cartesian coordinate system to trap optimum amount of solar energy

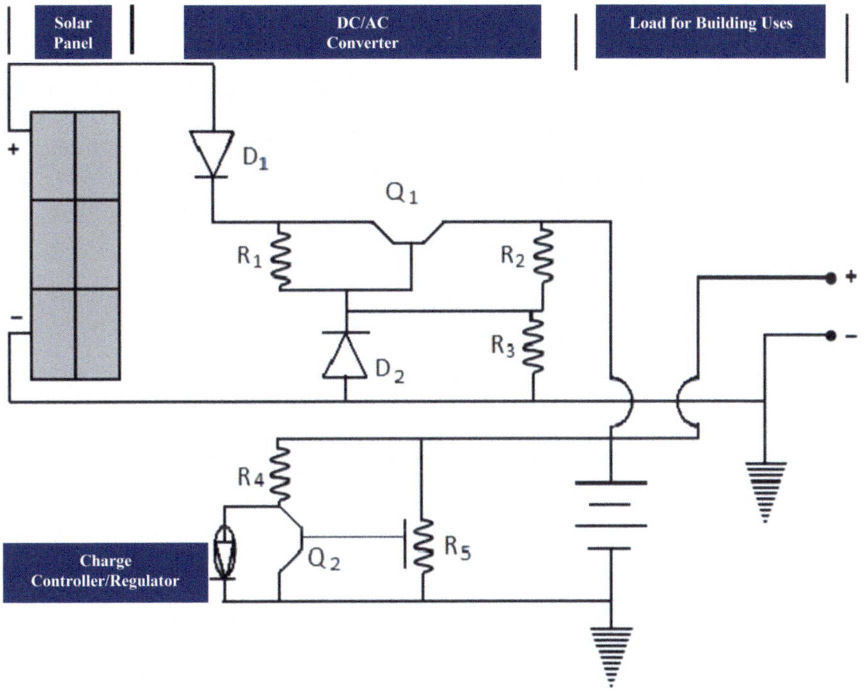

Fig. 11.2 The conceptual diagram shows the charge controller enables the printable solar cells to convert DC into AC to power the building subsequently

Hence, the photophysical response of a printable solar cell is responsible for capturing solar thermal energy, thus, the quantum electrodynamics has been analyzed using MATLAB for capturing solar energy considering the solar thermal conductivity and solar irradiance for printable solar cells [15, 19, 22]. Furthermore, the radiation energy density of the inner surface of the printable solar cell has been computed by employing the principles of modern quantum physics [14, 24, 49]. This calculation has considered the Maxwell–Boltzmann constant as well as electromagnetic wave statistics to establish a mathematical model of photovoltaic dynamics to achieve the highest possible efficiency of the printable solar cells mounted on the exterior curtain wall of a building system by using the following equation:

$$P_{pv} = \eta_{pvg} A_{pvg} G_t \tag{11.1}$$

In this equation, the η_{pvg} denotes the efficiency of a printable solar cell, A_{pvg} represents the area of the printable solar cell, and G_t is the solar radiation measured in watt meters on a certain module plane. Hence, this equation is further modified in a more specific manner as follows:

$$\eta_{pvg} = \eta_r \eta_{pc} \left[1 - \beta\left(T_c - T_{c\ ref}\right)\right] \tag{11.2}$$

Here, the temperature coefficient (β) has a range that lies between 0.004 and 0.0006, expressed in degree Celsius. The efficiency of the power conditioning module is represented as η_r, which stands for reference module efficiency and the temperature is measured in degree Celsius and is denoted by the symbol T_{cref}. Applying the following formula, the method can be utilized to determine the printable solar cell as

$$T_c = T_a + \left(\frac{NOCT - 20}{800}\right) G_t \tag{11.3}$$

Here, G_t is the solar irradiance expressed in Watts per square meter, and NOCT represents the temperature of the printable solar cell expressed in degree Celsius. Therefore, the net estimated solar irradiance in the printable solar cell by taking both the standard and diffuse solar irradiance into consideration, it can be expressed in the following equation:

$$I_t = I_b R_b + I_d R_d + (I_b + I_d) R_r \tag{11.4}$$

Hence, the printable solar cells are the semiconductor devices that generate energy by utilizing the standard diode incorporated to the cell with N_s series-connected arrays and N_p parallel-connected arrays, thus, the current generation related to the cell voltage can be expressed as following:

$$I = N_p \left[I_{ph} - I_{rs} \left[\exp\left(\frac{q(V + IR_s)}{AKTN_s} - 1 \right) \right] \right] \tag{11.5}$$

where

$$I_{rs} = I_{rr} \left(\frac{T}{T_r} \right)^3 exp \left[\frac{E_G}{AK} \left(\frac{1}{T_r} - \frac{1}{T} \right) \right] \tag{11.6}$$

In Eqs. 11.5 and 11.6, the parameter K represents Boltzmann's constant, the variable A denotes the idealist factor for diodes, and the variable T is the temperature of the cell in Kelvin. The charge of an electron, denoted by the symbol q, is 1.6×10^{-9} C. The reverse saturation current at absolute temperature (T) is denoted by the symbol I_{rs}, the cell reference temperature is denoted by T_r that is the reverse current, and I_{rr} represents band gap energy of the semiconductor that is utilized in the cell [23, 46, 48]. Thus, the photocurrent I_{ph} is being calculated as follows considering the temperature of the cell and the solar radiation rate:

$$I_{ph} = \left[I_{SCR} + k_i(T - T_r)\frac{S}{100}\right] \tag{11.7}$$

I_{SCR} denotes the short-circuit current of the printed solar cell under steady-state temperature and irradiance conditions. Here, the function of the scalar quantity of the solar irradiance is being measured in milliwatts per square centimeter [25, 44, 45]. The temperature coefficient of the short-circuit current is denoted by the symbol Ki. By combining the ideal diode with a single-diode model of the I–V characteristics, the following equation has been employed:

$$I = I_{ph} - I_D \tag{11.8}$$

$$I = I_{ph} - I_0\left[\exp\left(\frac{q(V + R_s I)}{AKT} - 1\right) - \frac{V + R_s I}{R_{sh}}\right] \tag{11.9}$$

Here, the photocurrent (I_{ph}) is defined as the diode current of I_D, the inverse current is I_0, the electron charge (q) is defined as 1.6×10^{-9} C, the series resistance is defined as R_s in Ohm, the cell voltage is defined as V, and thus, the current (I) is being as following to characterize the current printable solar cell:

$$I = I_{PV} - I_{D1} - I_{D2} - \left(\frac{V + IR_S}{R_{SH}}\right) \tag{11.10}$$

where

$$I_{D1} = I_{01}\left[\exp\left(\frac{V + IR_s}{a_1 V_{T1}}\right) - 1\right] \tag{11.11}$$

$$I_{D2} = I_{02}\left[\exp\left(\frac{V + IR_s}{a_2 V_{T2}}\right) - 1\right] \tag{11.12}$$

Subsequently, the thermal voltages V_{T1} and V_{T2} of the respective diodes are being measured considering the currents at diode 1 (D_1) and diode 2 (D_2). These currents are indicated as I_{01} and I_{02}, respectively. Thus, the voltage is being calculated as an idealist mode printable solar cell.

$$v_{oc} = \frac{V_{oc}}{cK\,T/q} \tag{11.13}$$

$$P_{\max} = \frac{\frac{V_{oc}}{cK\,T/q} - \ln\left(\frac{V_{oc}}{cK\,T/q} + 0.72\right)}{\left(1 + \frac{V_{oc}}{nK\,T/q}\right)} \left(1 - \frac{V_{oc}}{\frac{V_{oc}}{I_{SC}}}\right) \left(\frac{V_{oc0}}{1 + \beta \ln \frac{G_0}{G}}\right) \left(\frac{T_o}{T}\right)^y I_{sc0} \left(\frac{G}{G_o}\right)^a \tag{11.14}$$

Here, Eq. (11.14) reflects only the maximum amount of energy that can be produced by a single photovoltaic printable solar cell module, despite the fact that a practical system will make use of a large number of printable solar cells that are connected in series and parallel [26, 36, 37]. For this reason, the printable solar cells can be characterized using the equation that reflects the total power generation for an array of Ns cells linked in series and Np cells linked in parallel, in addition to the power factor (P_M) for each module as follows:

$$P_{\text{array}} = N_s N_p P_M \tag{11.15}$$

Consequently, the photophysical processes will be activated in the photon flux in printable solar cells to transform it into the energy level charges. Considering the simulation of the solar radiation energy density by using classical statistical physics, the maximum solar energy generation for a single photon can be calculated at a rate of 1.4 eV that is equivalent to 27.77 MW/m^2 eV [16, 41, 42].

Results and Discussion

The activation of the printable solar cells is defined by the application of the current–voltage (I–V) equation using single-diode mode that governs the I–V relationship as follows:

$$\mathrm{I} = \mathrm{IL} - \mathrm{IO}\left\{\exp^{\left[\frac{q(V+I_{RS})}{AkTc}\right]} - 1\right\} - \frac{(V + I_{RS})}{R_{Sh}} \tag{11.16}$$

Here, I_L stands for current generation, b stands for current of the diode, R_s stands for series resistance, A stands for the diode's passive function, k is Boltzmann's constant, q is electron charge amplitude, and T_C is cell temperature. Subsequently, the I–q relationship in printable solar cells can be represented as follows:

$$\mathrm{I}_O = \mathrm{I}_{\mathrm{RS}} \left(\frac{T_C}{T_{ref}}\right)^3 \exp\left[\frac{qEG\left(\frac{1}{T_{ref}} - \frac{1}{T_C}\right)}{KA}\right] \tag{11.17}$$

and

$$V_{mo} = -I_{mo}R_{Smo} + K_{mo}\log\left(\frac{I_{Lmo} - I_{mo} + I_{Omo}}{I_{Omo}}\right) \tag{11.18}$$

In this context, the solar cell-generated I_{mo} is entering the diode I_{Omo} in the series resistance of R_{smo}, where the factorial constant is K_{mo}. To determine the series resistance, $R_{smo} = NS \times RS$, and the constant factor, $K_{mo} = NS \times K$, the circuit is being linked to all N_S cells in conjunction with one another in series [27, 39, 40]. The series-connected cells experience a particular quantity of current flow that can be addressed by the formulation of the module for the NS series of connected cells as follows:

$$V_{mo} = -I_{mo}N_SR_S + N_SK\log\left(\frac{I_L - I_{mo} + I_o}{I_o}\right) \tag{11.19}$$

Once all of the NP cells are linked in parallel circuit, the computation current and voltage for the parallel connection can be represented as [28, 29, 38]

$$V_{mo} = -I_{mo}\frac{Rs}{Np} + K\log\left(\frac{N_{sh}I_L - I_{mo} + N_pI_o}{N_pI_o}\right) \tag{11.20}$$

Hence, this printable solar cell-generated current, which is emitted from the solar irradiance, is being calculated considering the following equation:

$$L = G[I_{SC} + K_I(T_C - T_{ref})] * V_{mo} \tag{11.21}$$

where I_{sc} denotes printable solar cell current, K_I denotes the printable solar cell coefficient factor, T_{ref} denotes the printable solar cell's temperature, and G denotes the solar energy trapping rate in KW/m^2.

Conversion of Electricity

Subsequently, an electrical circuit consisting of a single diode has been utilized to successfully convert solar energy into electricity by employing a thin semiconductor disk that is hooked into a silicon film attached to printable solar cells. This circuit obtains the open circuit voltage, V_{oc}, which in turn defines the optimum power point, which can be accomplished by determining the optimum current and voltage immediately following the capture of the solar irradiance in the cells [30, 33]. Considering this, the mode of circuit (I_{mp}, V_{mp}) is extremely important for achieving the highest possible power output from the printable solar cells [27, 31]. By conducting a precise analysis of the circuit, the printable solar cell has been demonstrated accurately to the cell that is capable of producing the highest energy through the flow of current and voltage (*see* Fig. 11.3).

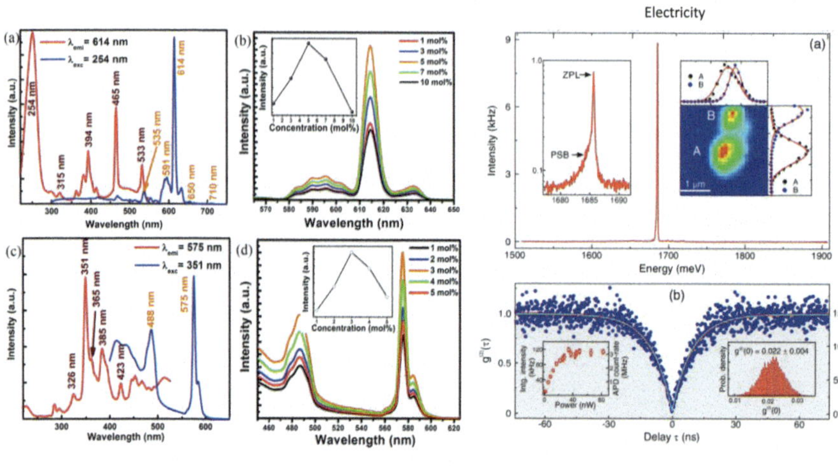

(A) **(B)**

Fig. 11.3 (**a**) The intensity of solar energy production is measured at a number of different wavelengths. (**b**) The current–voltage characteristics of printable solar cells to determine the theoretical role that the net ideal energy output intensity plays in relation to the delay caused by the coincidence factor

Simply, this research revealed the relationship between the amount of steady-state light-quanta solar energy and the efficiency of printable solar cell in converting solar irradiance into electrical energy. Hence, this work suggested that the correlation between the current and the voltage in printable solar cells is related to the density of light-quanta emission inside the cells. This density is a measurement of a specific sort of polarization of light quanta that is achieved by the optimal solar radiation with an energy value of 1.4 eV which is equivalent to 27.77 MW/m^2 considering average peak solar irradiance for 5 h in a day, which is comparable to 27,770 kW/year or 7.6 kW/day [28, 34]. More precisely, it can describe that the process of converting solar energy into direct current (DC) power and then into alternating current (AC) power results in "derating factor," is 0.8. Hence, the current conversion by printable solar cells is 125% with an efficiency of 0.8 of solar cells is (27,770 × 1.25 × 0.8) = 27,770 kW/year or 7.6 kW/day. A standard 5000 SF residential house requires an average of 7 kW/day. Thus, the energy generated by the printable solar cells is equivalent to 7.6 kW/day, which in fact will meet the net energy of a house by installing only 1 m^2 printable solar cells on the exterior curtain wall. Subsequently, the average energy consumption rate monthly of commercial office or buildings of a foot-printing $100'_{\text{length}} \times 100'_{\text{width}}$ with 100′ tall (10 floors) will require only 20 m^2 exterior curtain wall to be layered up by printable solar cells to meet its net energy of 140 kW/day (Table 11.1).

Table 11.1 Cost of the printable solar cells for energy production of a 10 stories building

Items	Materials Cost	Labor Cost	Equipment Cost	GC and OH Cost	Total Cost
Printable solar cells	$11,000	$5100	$2600	$3600	$21,000
Instruments	$2100	$1100	$2100	$1100	$6000
Control system	$2600	$1100	$1100	$1000	$5400
Thirty years cost at $0.05/kwh/ month for 4000 kwh					$72,000
				Net Cost	**$105,400**

Savings on Energy Cost

As the net cost of 30 years of energy purchase from a conventional utility source from a standard industry is $172,800 (considering 0.12/kwh of 4000 kW/month) in comparison to the printable solar cell energy cost for 30 years of $104,400, the net cost savings of $67,400 indeed would represent significant economic value for energy consumption in addition to the nature of renewable and environmentally friendly characteristics of this energy source.

Conclusions

There has been an increasing level of concern regarding the severe threat associated with the presence of greenhouse gases (GHGs) in the atmosphere. It is, therefore, an urgent need for the intelligent use of solar energy as conventional energy consumption is a primary factor in environmental challenges and climate change. Furthermore, the finite level of the fossil fuel reserve presents a significant obstacle for forthcoming energy provision, given that the utilization of fossil fuels is projected to cease during the subsequent decades. Interestingly, this study suggested that innovative energy technology and its intellectual means of application through the induction of the printable solar cells on the outer layer of the curtain wall considering both economically viable and environmentally reliable will indeed mitigate the net energy demand in the building sector throughout the world which will reduce 40% of conventional energy consumption in comparison to the total energy in every sector globally. Simply, the installation of printable solar cells at the outside curtain walls for capturing solar energy as a DC and then converting it into AC to meet the required energy demand for that building shall indeed an excellent technology in energy and building science to mitigate energy and climate crisis significantly.

Acknowledgments The author declares that this research chapter does not have any conflict of interest for publishing in any suitable publisher.

References

1. Armani, D. K., Kippenberg, T. J., Spillane, S. M., & Vahala, K. J. (2003). Ultra-high-Q toroid microcavity on a chip. *Nature, 421*, 925.
2. Artemyev, N., Jentschura, U. D., Serbo, V. G., & Surzhykov, A. (2012). Strong electromagnetic field EFFECTS in ultra-relativistic heavy-ion collisions. *European Physical Journal C, 72*, 1935.
3. Birnbaum, K. M., et al. (2005). Photon blockade in an optical cavity with one trapped atom. *Nature, 436*, 87–90.
4. Busch, K., von Freymann, G., Linden, S., Mingaleev, S. F., Tkeshelashvili, L., & Wegener, M. (2007). Periodic nanostructures for photonics. *Physics Reports, 444*, 101.
5. Chang, D. E., Sørensen, A. S., Demler, E. A., & Lukin, M. D. (2007). A single-photon transistor using nanoscale surface plasmons. *Nature Physics, 3*, 807–812.
6. Chen, J., Wang, C., Zhang, R., & Xiao, J. (2012). Multiple plasmon-induced transparencies in coupled-resonator systems. *Optics Letters, 37*, 5133–5135.
7. Cheng, M., & Song, Y. (2012). Fano resonance analysis in a pair of semiconductor quantum dots coupling to a metal nanowire. *Optics Letters, 37*, 978–980.
8. Dayan, B., et al. (2008). A photon turnstile dynamically regulated by one atom. *Science, 319*, 1062–1065.
9. Douglas, J. S., Habibian, H., Hung, C., Gorshkov, A., & Kimble, H. (2015). *and Chang*. D. Quantum many-body models with cold atoms coupled to photonic crystals.
10. Eichler, J., & Stöhlker, T. (2007). Radiative electron capture in relativistic ion-atom collisions and the photoelectric effect in hydrogen-like high-Z systems. *Physics Reports, 439*, 1.
11. Englund, D., et al. (2010). Resonant excitation of a quantum dot strongly coupled to a photonic crystal nanocavity. *Physical Review Letters, 104*, 073904.
12. Gleyzes, S., et al. (2007). Quantum jumps of light recording the birth and death of a photon in a cavity. *Nature, 446*, 297.
13. Gould, R. J. (1967). Pair production in photon-photon collisions. *Physical Review*.
14. Guerlin, C., et al. (2007). Progressive field-state collapse and quantum non-demolition photon counting. *Nature, 448*, 889.
15. Han, Z., & Bozhevolnyi, S. I. (2011). Plasmon-induced transparency with detuned ultracompact Fabry-Pérot resonators in integrated plasmonic devices. *Optics Express, 19*, 3251–3257.
16. Hencken, K. (2006). Transverse momentum distribution of vector mesons produced in ultraperipheral relativistic heavy ion collisions. *Physical Review Letters*.
17. Hossain, M. F. (2016). Solar energy integration into advanced building design for meeting energy demand. *International Journal of Energy Research, 40*, 1293–1300.
18. Hossain, M. F. (2017). Design and construction of ultra-relativistic collision PV panel and its application into building sector to mitigate total energy demand. *Journal of Building Engineering*. https://doi.org/10.1016/j.jobe.2016.12.005
19. Hossain, M. F. (2017). Green science: Independent building technology to mitigate energy, environment, and climate change. *Renewable and Sustainable Energy Reviews*. https://doi.org/10.1016/j.rser.2017.01.136
20. Hossain, M. F. (2018). Green Science: Advanced Building Design Technology to Mitigate Energy and Environment. *Renewable and Sustainable Energy Reviews., 81*(2), 3051–3060.
21. Hossain, M. F. (2018). Photonic thermal energy control to naturally cool and heat the building. *Advanced Thermal Engineering, 131*, 576–586.
22. Hossain, M. F. (2018). Transforming dark photon into sustainable energy. *International Journal of Energy and Environmental Engineering*. https://doi.org/10.1007/s40095-017-0257-1
23. Huang, Y., Min, C., & Veronis, G. (2011). Subwavelength slow-light waveguides based on a plasmonic analogue of electromagnetically induced transparency. *Applied Physics Letters, 99*, 143117.

24. Huang, J. F., Shi, T., Sun, C. P., & Nori, F. (2013). Controlling single-photon transport in waveguides with finite cross section. *Physical Review A, 88*, 013836.
25. Jentschura, U., Hencken, K., & Serbo, V. (2008). Revisiting unitarity corrections for electromagnetic processes in collisions of relativistic nuclei. *The European Physical Journal C, 58*(2), 281–289.
26. Joannopoulos, J. D., Villeneuve, P. R., & Fan, S. (1997). Photnic crystals: Putting a new twist on light. *Nature, 386*, 143.
27. Klein, S. A. (1975). Calculation of flat-plate collector loss coefficients. *Solar Energy, 17*, 79–80.
28. Kofman, A. G., Kurizki, G., & Sherman, B. (1994). Spontaneous and induced atomic decay in photonic band structures. *Journal of Modern Optics, 41*, 353.
29. Kolchin, P., Oulton, R. F., & Zhang, X. (2011). Nonlinear quantum optics in a waveguide: Distinct single photons strongly interacting at the single atom level. *Physical Review Letters, 106*, 113601.
30. Lang, C., et al. (2011). Observation of resonant photon blockade at microwave frequencies using correlation function measurements. *Physical Review Letters, 106*, 243601.
31. Lei, C. U., & Zhang, W. M. (2012). A quantum photonic dissipative transport theory. *Annals of Physics, 327*, 1408.
32. Li, Q., Xu, D. Z., Cai, C. Y., & Sun, C. P. (2013). Recoil effects of a motional scatterer on single-photon scattering in one dimension. *Scientific Reports.*
33. Liao, J. Q., & Law, C. K. (2010). Correlated two-photon transport in a one-dimensional waveguide side-coupled to a nonlinear cavity. *Physical Review A, 82*, 053836.
34. Liao, J. Q., & Law, C. K. (2013). Correlated two-photon scattering in cavity optomechanics. *Physical Review A, 87*, 043809.
35. Lo, P., Xiong, H., & Zhang, W. (2015). Breakdown of Bose-Einstein distribution in photonic crystals. *Scientific Reports.*
36. Najjari, B., Voitkiv, A., Artemyev, A., & Surzhykov, A. (2009). Simultaneous electron capture and bound-free pair production in relativistic collisions of heavy nuclei with atoms. *Physical Review A, 80*, 012701.
37. O'Shea, D., Junge, C., Volz, J., & Rauschenbeutel, A. (2013). Fiber-optical switch controlled by a single atom. *Physical Review Letters, 111*, 193601.
38. Reinhard, A. (2011). Strongly correlated photons on a chip. *Nature Photonics.*
39. Roy, D. (2013). Two-photon scattering of a tightly focused weak light beam from a small atomic ensemble: An optical probe to detect atomic level structures. *Physical Review A, 87*, 063819.
40. Ruiz, A. (2014). Partial recovery of a potential from backscattering data. In *Communications in partial differential equations* (Springer tracts in modern physics). Springer.
41. Sánchez Muñoz, C., Laussy, F., Valle, E., Tejedor, C., & González-Tudela, A. (2018). Filtering multiphoton emission from state-of-the-art cavity quantum electrodynamics. *Optica, 5*(1), 14–26.
42. Sayrin, C., et al. (2011). Real-time quantum feedback prepares and stabilizes photon number states. *Nature, 477*, 73.
43. Shen, J. T., & Fan, S. (2007). Strongly correlated two-photon transport in a one-dimensional waveguide coupled to a two-level system. *Physical Review Letters, 98*, 153003.
44. Shi, T., Fan, S., & Sun, C. P. (2011). Two-photon transport in a waveguide coupled to a cavity in a two-level system. *Physical Review A, 84*, 063803.
45. Tame, M. S., McEnery, K. R., Özdemir, Ş. K., Lee, J., Maier, S. A., & Kim, M. S. (2013). Quantum plasmonics. *Nature Physics.*
46. Valluri, S. R., Becker, U., Grun, N., & Scheid, W. (1984). Relativistic collisions of highly-charged ions. *Journal of Physics B: Atomic, Molecular and Optical Physics, 17*, 4359.

47. Wang, Y., Zhang, Y., Zhang, Q., Zou, B., & Schwingenschlogl, U. (2016). Dynamics of single photon transport in a one-dimensional waveguide twopoint coupled with a Jaynes-Cummings system. *Scientific Reports*.
48. Yang, L., Wang, S., Zeng, Q., Zhang, Z., Pei, T., Li, Y., & Peng, L. (2011). Efficient photovoltage multiplication in carbon nanotubes. *Nature Photonics*, 672–676.
49. Zhang, W. M., Lo, P. Y., Xiong, H. N., Tu, M. W. Y., & Nori, F. (2012). General non-markovian dynamics of open quantum systems. *Physical Review Letters, 109*, 170402.

Chapter 12
Implementation of Sonic Reaction Technology in Building Design to Eliminate Pathogens Naturally

Introduction

In an occupied room measuring 10 by 10 ft, the typical number of microorganisms found in an indoor environment is approximately 8000 different species of viruses, bacteria, molds, and fungus, with the concentrations reaching 10^5 pathogens per cubic meter [2, 6, 14]. Infectious microorganisms, often known as pathogens, include bacterium, protozoan, prion, viroid, fungus, and virus, including coronavirus disease 2019 (COVID-19). These pathogens are hazardous beings as they can be produced by these microscopic organisms, which can easily invade the human body by absorbing, adsorbing, breathing, swallowing, and form toxins inside the human bodies that infiltrate the tissues, hijack nutrition, immunosuppress the human body, and cause significant sickness or death to the humans [1, 7]. Hence, the fatal infectious disease COVID-19 presence inside the buildings was responsible for the severe atypical respiratory syndrome (SARS-CoV-2) disease globally, and thus, the World Health Organization (WHO) has stated that "the 2019–2020 coronavirus outbreak" is a pandemic [11, 27, 29]. Aside from the COVID-19 pandemic catastrophe, the world is not equipped to combat any future dangerous outbreaks that might be significantly more lethal than COVID-19, which is likely to wipe out the whole human race off the face of the earth. Some pathogens have the incredibly smart capacity to modify their biochemical indie the human body to deceive any vaccines that are intended to kill them. Therefore, the development of preventative technologies is an urgent necessity to eradicate these lethal pathogens before they penetrate the human body. It is interesting that Hossain ultrasonic germicidal frequency (*HuGF*) possesses the exceptional functionality to eradicate these lethal pathogens in a matter of seconds, regardless of their ability to change their metabolism pathways. Consequently, the ultrasonic wavelength structure of the HuGF has been investigated in this research, taking into consideration the process by which it is formed by the inside wall of the building. Then, this HuGF technology is being employed to release high-frequency wavelengths ranging from 10 kHz to 1 mHz,

© The Author(s), under exclusive license to Springer Nature Switzerland AG 2025

M. F. Hossain, *Sustainable Global Development in Environment, Energy, Building, and Infrastructure*, https://doi.org/10.1007/978-3-031-84429-4_12

which are capable of destroying the nucleic acids and DNA and/or RNA structure of pathogens. To put it more simply, the ultrasonic response technology should be implemented to construct all buildings with the capability of producing HuGF to eliminate all pathogens inside the building naturally before they can enter the human body.

Materials and Methods

HuGF Formation by the Interior Building Wall

The frequency field of the sound through the design of the inside wall of the building has been clarification to release HuGF in the range of 10 kHz to 1 mHz for a limited period of time when there are no people living inside the structure [3, 12]. Naturally, the sound frequency generation has been modeled taking into consideration the areal parameter of the building's interior wall and expressed by the following Hamiltonian's equation:

$$H = \sum \omega_{ci} a_i^\dagger a_i + \sum_K \omega_k b_k^\dagger b_k + \sum_{ik} \left(V_{ik} a^\dagger b_k + V_{ik}^* b_k^\dagger a_i \right) \tag{12.1}$$

where $a_i\left(a_i^\dagger\right)$ represents the kinematics of the sound frequency, $b_k^\dagger b_k$ denotes the kinematics of the sound, and $a^\dagger b_k$ denotes the cofactors that indicate the mode of sound release rate inside the building [4, 15]. Subsequently, the entire sound frequency spectrum is being calculated undertaking into account the inside wall of the building and expressed by the following equation [13, 16, 18]:

$$\begin{aligned} \rho(t) = & -i\left[H_c'(t), \rho(t)\right] + \sum_{ij} \left\{ k_{ij}(t) \left[2a_j \rho(t) a_i^\dagger - a_i^\dagger a_j \rho(t) - \rho(t) a_i^\dagger a_j \right] \right. \\ & \left. + k_{ij}(t) \left[a_i^\dagger \rho(t) a_j + a_j \rho(t) a_i^\dagger - a_i^\dagger a_j \rho(t) - \rho(t) a_j a_i^\dagger \right] \right\} \end{aligned} \tag{12.2}$$

In this context, the sound frequency is being factored considering the time-related function and then being employed by integration of vector perturbative principle and thus, the equation has been rewritten as follows:

$$\omega_c' (t) = -Im[u (t, t_0)/u (t, t_0)] \tag{12.3}$$

$$k (t) = - Re \ [u (t, t_0)/u (t, t_0)] \tag{12.4}$$

$$\widehat{k}(t) = \dot{v} (t, t) + 2v (t, t) \ k(t) \tag{12.5}$$

It indicates the sound frequency area and the sound frequency proliferation within the building. Thus, this equation is modified further by employing the equilibrium sound frequency dynamic mechanism, which is as follows:

$$\dot{u}\,(t,t_0) = -i\omega_c u\,(t,t_0) - \int_{t_0}^{t} dt' g\,(t-t')\,u\,(t',t_0) \tag{12.6}$$

$$v\,(t,t) = \int_{t_0}^{t} dt \int_{0}^{t} dt_2\;\; u^*\,(t_1,t_0)\widehat{g}\,(t_1-t_2)\,u\,(t_2,t_0) \tag{12.7}$$

Here, the scalar functions in equations are denoted by the symbol vc, which stands for the fundamental frequency of the sound that is being clarified by the scalar function of the sound frequency. Hence, the amount of sound frequency created inside the building per unit area is being considered the density of state (DOS) sound frequency generation (***HuGF***) which can be expressed as follows:

$$J\,(\omega) = \sum_k |V_k|^2\,\delta\,(\omega-\omega_k) = \varrho\,(\omega)\mid V\,(\omega)|^2 = [n*e(1+2n)]^4 \tag{12.8}$$

Killing Pathogens

Owing to the fact that the sound frequencies are being transmitted through vector fields, it is thus, inevitable as these frequencies will react with other sound frequencies, which will then interact with other sound frequency momentum [10, 19]. Consequently, it will be possible to release a significantly greater amount of ultrasound frequency HuGF continuously without any disturbances [22, 26, 36]. As a result of the fact that ultrasonic sound will have a high frequency of sound wave HuGF, this wave will possess sufficient energy to break the biochemical bonds that are present in pathogens, which will result in significant damage to the structure of pathogenic cells. Subsequently, the sonic-Polymerase chain reaction (PCR) has been implemented to determine the effects of ultrasonic sound waves on pathogens, including on the corona virus (COVID-19). Hence, the formation of HuGF from the interior of the building is being controlled by a computerized *soundometer* to control the sound wavelength frequency that works on pathogens.

This computerized analysis of the sound frequency was conducted to verify and confirm the required sound frequency to break down the molecular bonds of microorganisms' DNA and/or RNA and/or thymine dimers (*see* Fig. 12.1).

Once the HuGF is released within the building, it will render pathogens inactive by destroying the deoxyribonucleic acid (DNA) and ribonucleic acid (RNA) structures that they possess. For the simple reason that when the DNA and/or RNA of the pathogens are struck by the HuGF, it will generate a very powerful sound force that will cause the dimers, which are covalent bonds between the identical nucleic acids, of the pathogens to be broken down. The failure of dimers will, therefore, result in a

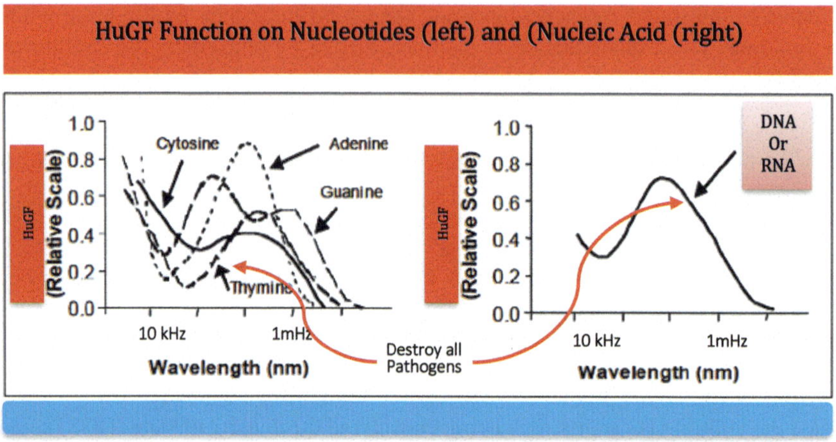

Fig. 12.1 The high-frequency sound application shows that once ***HuGF*** sound of 10 kHz to 1 mHz is implemented inside the building, it will start to disinfect all pathogens immediately

barrier in the transcription of information of DNA and/or RNA, which will, in turn, result in the breakdown of the biochemical function and replication of pathogens, causing them to die quickly.

Results and Discussion

HuGF Formation by the Interior Building Wall

The sound frequency dynamics have been estimated taking into consideration the inner unit area of the building [17, 34, 37]. This computation was conducted to quantitatively confirm that the interior wall of the building has the capability for the creation of HuGF. Because the unit area has a consistent sound frequency release rate, the creation of HuGF will have enough frequency dynamic modes that can be either one-dimensional (1D), two-dimensional (2D), or three-dimensional (3D) ultrasound frequencies within the building [5, 8, 35]. Subsequently, a high sound frequency cutting-off is also being estimated from the bifurcation of the density of state (DOS) of sound in a three-dimensional directional factor to control the pace of the high-frequency sound waves that are produced inside the building [9, 21]. After that, a sharpened sound of high-level frequency is utilized to keep the whole DOS in three dimensions, two dimensions, and one dimension within the building. Consequently, the net sound DOS requirement inside the building is being precisely calculated by controlling the sound frequency cutting-off considering the DOS inside the building [20, 23]. The determination of this sound DOS is, therefore, accomplished by ng three-dimensional, two-dimensional, and one-dimensional analyses of the sound frequency that will verify the necessary qualitative state of sound frequency for disrupting the pathogen's biochemical activities [24, 32, 33]. It is, therefore, suggested that the required sound frequency is accurate by the implementation of the pure algorithm of sound frequency into the combined 3D, 2D, and 1D, thus, it was expressed as follows:

$$J(\omega) = \varrho(\omega)|V(\omega)|^2 \tag{12.10}$$

In this context, it has been taken into consideration as the sound frequency and the dynamics of the sound frequency are the function of the UVGI proliferation and thus, Eq. 12.10 can be detailed as

$$u(t, t_0) = \frac{1}{1 - \Sigma'(\omega_b)} e^{-i\omega(t - t_0)} + \int_{\omega_e}^{\infty} d\omega \frac{J(\omega)e^{-i\omega(t - t_0)}}{[\omega - \omega_c - \Delta(\omega)]^2 + \pi^2 J^2(\omega)} \tag{12.11}$$

where $\Sigma'(\omega_b) = [\partial\Sigma(\omega)/\partial\omega]_{\omega = \omega_b}$ and $\Sigma(\omega)$ denote the sound frequency induction,

$$\Sigma(\omega) = \int_{\omega_e}^{\infty} d\omega' \frac{J(\omega')}{\omega - \omega'} \tag{12.12}$$

Here, the frequency ω_b in Eq. (12.2) represents the ***HuGF*** frequency mode as $(0 < \omega_b < \omega_e)$ and is calculated in a simplified way as $\omega_b - \omega_c - \Delta(\omega_b) = 0$, where $\lesssim\Delta(\omega) = \mathcal{P}\left[\int d\omega' \frac{\mathrm{J}(\omega')}{\omega - \omega'}\right]$ is a principal value integral.

Consequently, the appropriate HuGF sound frequency has been computed inside the building, taking into consideration the release magnitude of sound, while also taking into consideration the variable detuning integration of sound frequency [25, 30]. Interestingly, this sound frequency deliberation rate demonstrated that sound frequency HuGF is released at an equilibrium rate with a range of 10 kHz to 1 MHz inside the building. This finding indicated that HuGF release rate inside the building is practically in a dynamic mode, and thus, it can be expressed as

$$v(t, t) = \int_{t_0}^{t} dt_1 \int_{t_0}^{t} dt_2 u^*(t_1, t_0)\tilde{g}(t_1, t_2)u(t_2, t_0) \tag{12.13}$$

Here, the sound frequency $\tilde{g}(t_1, t_2) = \int d\omega J(\omega)\overline{n}(\omega, T)e^{-i\omega(t - t')}$ is related to the ***HuGF*** release rate, where $\overline{n}(\omega, T) = 1/\left[e^{\hbar\omega/k_B T} - 1\right]$ is the generation of the sonic wave inside the building at the time of T and it is expressed as

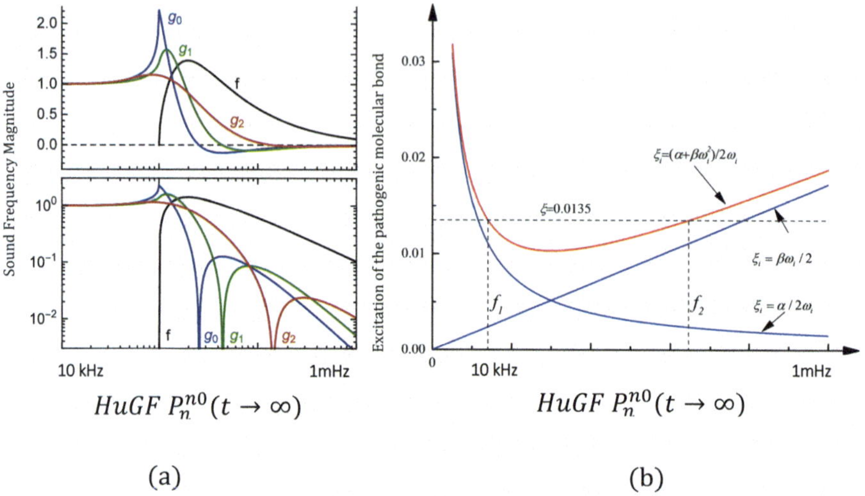

Fig. 12.2 (**a**) The creation of HuFG at three different wavelengths from the inside wall of the building. (**b**) HuFG interaction taking into consideration its frequency. As the frequency of the HuGF rises, the pathogenic bond becomes twisted that causes damage to the DNA and/or RNA

$$v(t, t \to \infty) = \int_{\omega_e}^{\infty} d\omega \mathcal{V}(\omega)$$

with

$$\mathcal{V}(\omega) = \overline{n}(\omega, T)[\mathcal{D}_l(\omega) + \mathcal{D}_d(\omega)] \tag{12.14}$$

Simply, this finding demonstrated that the sound frequency at this location has an extremely high velocity of wave at the interior of the building. Therefore, this HuGF will have an extremely high force at the steady-state limit, which will quickly strike the biochemical link of the pathogens (*see* Fig. 12.2).

Killing Pathogens

The HuGF, the frequency of sound, has been clarified as it has the extreme velocity of wave force, and thus, this force will be linearly polarized inside the building at a sinusoidal wave frequency [28, 31]. Then, it will be scattering inside the building at direction +z through a homogeneous, isotropic, dissipation vector of wave with the frequency between 10 kHz and 1 mHz. This wave of the frequency has the outstanding capability to break the DNA and/or RNA of the fatal pathogens in seconds regardless of their pathogens' frequent mechanism of mutation. In a

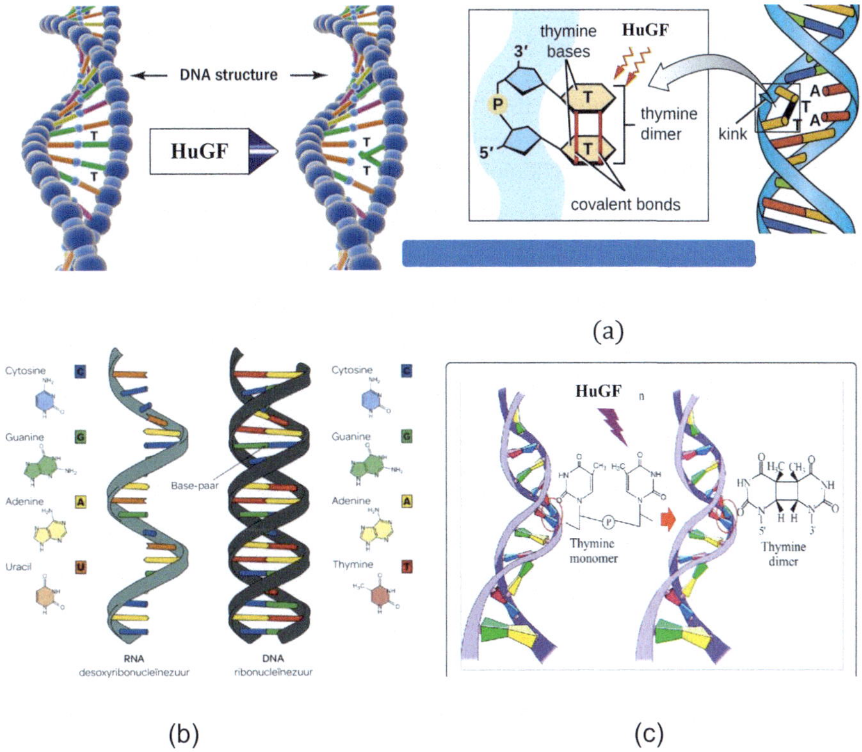

Fig. 12.3 Demonstrating the HuGF influences on (**a**) the structure of DNA; (**b**) on the DNA and RNA bonds of the pathogens; and (**c**) disinfection of the thymine dimer of the pathogens to ultimately disable their biochemical function and reproduction

nutshell, it will destroy all biochemical bonds that are composed of the nucleotides adenine (A), thymine (T), guanine (G), and cytosine (C) that are chemically coded with its DNA and/or RNA structures (*see* Fig. 12.3). Subsequently, it will completely damage their cyclobutene–pyrimidine–dimers (CPD). This will result eventually, the pyrimidine dimers will cause a kink into the DNA and/or RNA strands, which will stop transcription and protein synthesis will result in severe damage to cellular function (*see* Fig. 12.3). Then, these pathogens are to be rendered incapable of performing any of their biological function, and as a result, they will die in a second.

Conclusions

Advanced building design technology has been proposed to generate *Hossain ultrasonic germicidal frequency (**HuGF**)* inside the building in the denaturation of all pathogens, including COVID-19. This is because the HuGF has the excellent capability to damage the DNA and/or RNA structure will result in all thymine dimers as well as the biochemical bond of the pathogens broken down completely. As a result, pathogens including COVID-19 will no longer be able to proliferate their biochemical function, which will make them to die eventually. Concretely, the utilization of *Hossain ultrasonic germicidal frequency (**HuGF**)*, in the context of advanced building design technology, which will entail the formation of high-frequency wavelengths ranging from 10 kHz to 1 mHz, will eliminate all pathogens inside the building by causing damage to the DNA and/or RNA structures of all pathogens. Indeed, this would be the most promising field of science that has the potential to eliminate all pathogens inside the building before they can manifest themselves in the human body.

Acknowledgments The author confirms that this research does have any conflicts of interest and bears no restriction to get published a notable publishing company.

References

1. Armani, D. K., Kippenberg, T. J., Spillane, S. M., & Vahala, K. J. (2003). Ultra-high-Q toroid microcavity on a chip. *Nature, 421*, 925.
2. Birnbaum, K. M., et al. (2005). Photon blockade in an optical cavity with one trapped atom. *Nature, 436*, 87–90.
3. Busch, K., von Freymann, G., Linden, S., Mingaleev, S. F., Tkeshelashvili, L., & Wegener, M. (2007). Periodic nanostructures for photonics. *Physics Reports, 444*, 101.
4. Chang, D. E., Sørensen, A. S., Demler, E. A., & Lukin, M. D. (2007). A single-photon transistor using nanoscale surface plasmons. *Nature Physics, 3*, 807–812.
5. Chen, J., Wang, C., Zhang, R., & Xiao, J. (2012). Multiple plasmon-induced transparencies in coupled-resonator systems. *Optics Letters, 37*, 5133–5135.
6. Cheng, M. T., & Song, Y. Y. (2012). Fano resonance analysis in a pair of semiconductor quantum dots coupling to a metal nanowire. *Optics Letters, 37*, 978–980.
7. Dayan, B., et al. (2008). A photon turnstile dynamically regulated by one atom. *Science, 319*, 1062–1065.
8. Douglas, J. S., Habibian, H., Hung, C.-L., Gorshkov, A. V., Kimble, H. J., & Chang, D. E. (2015). Quantum many-body models with cold atoms coupled to photonic crystals. *Nature Photonics*.
9. Englund, D., et al. (2010). Resonant excitation of a quantum dot strongly coupled to a photonic crystal nanocavity. *Physical Review Letters, 104*, 073904.
10. Gleyzes, S., et al. (2007). Quantum jumps of light recording the birth and death of a photon in a cavity. *Nature, 446*, 297.
11. Guerlin, C., et al. (2007). Progressive field-state collapse and quantum non-demolition photon counting. *Nature, 448*, 889.
12. Hossain, M. F. (2016). Solar energy integration into advanced building design for meeting energy demand and environment problem. *International Journal of Energy Research*.
13. Hossain, M. F. (2017). Green science: Independent building technology to mitigate energy, environment, and climate change. *Renewable and Sustainable Energy Reviews*. https://doi.org/10.1016/j.rser.2017.01.136
14. Hossain, M. F. (2017). Design and construction of ultra-relativistic collision PV panel and its application into building sector to mitigate Total energy demand. *Journal of Building Engineering*. https://doi.org/10.1016/j.jobe.2016.12.005

15. Hossain, M. F. (2018). Photon energy amplification for the design of a micro PV panel. *International Journal of Energy Research.*
16. Hossain, M. F. (2019). Sustainable technology for energy and environmental benign building design. *Journal of Building Engineering.*
17. Hossain, M., & Faruque. (2018). Transforming dark photons into sustainable energy. *International Journal of Energy and Environmental Engineering.*
18. Hossain, M., & Faruque. (2018). Green science: Advanced building design technology to mitigate energy and environment. *Renewable and Sustainable Energy Reviews.*
19. Hou, J., Wang, H., & Liu, P. (2018). Applying the blockchain technology to promote the development of distributed photovoltaic in China. *International Journal of Energy Research.*
20. Huang, Y., Min, C., & Veronis, G. (2011). Subwavelength slow-light waveguides based on a plasmonic analogue of electromagnetically induced transparency. *Applied Physics Letters, 99*, 143117.
21. Huang, J. F., Shi, T., Sun, C. P., & Nori, F. (2013). Controlling single-photon transport in waveguides with finite cross section. *Physical Review A, 88*, 013836.
22. Joannopoulos, J. D., Villeneuve, P. R., & Fan, S. (1997). Photnic crystals: Putting a new twist on light. *Nature, 386*, 143.
23. John, S., & Wang, J. (1991). Quantum optics of localized light in a photonic band gap. *Physical Review B, 43*, 12772.
24. Kolchin, P., Oulton, R. F., & Zhang, X. (2011). Nonlinear quantum optics in a waveguide: Distinct single photons strongly interacting at the single atom level. *Physical Review Letters, 106*, 113601.
25. Lei, C. U., & Zhang, W. M. (2012). A quantum photonic dissipative transport theory. *Annals of Physics, 327*, 1408.
26. Liao, J. Q., & Law, C. K. (2010). Correlated two-photon transport in a one-dimensional waveguide side-coupled to a nonlinear cavity. *Physical Review A, 82*, 053836.
27. Longo, P., Schmitteckert, P., & Busch, K. (2011). Few-photon transport in low-dimensional systems. *Physical Review A, 83*, 063828.
28. O'Shea, D., Junge, C., Volz, J., & Rauschenbeutel, A. (2013). Fiber-optical switch controlled by a single atom. *Physical Review Letters, 111*, 193601.
29. Reinhard, A. (2011). Strongly correlated photons on a chip. *Nature Photonics.*
30. Roy, D. (2013). Two-photon scattering of a tightly focused weak light beam from a small atomic ensemble: An optical probe to detect atomic level structures. *Physical Review A, 87*, 063819.
31. Saloux, E. (2011). Explicit model of photovoltaic panels to determine voltages and currents at the maximum power point. *Solar Energy.*
32. Sayrin, C., et al. (2011). Real-time quantum feedback prepares and stabilizes photon number states. *Nature, 477*, 73.
33. Shen, J. T., & Fan, S. (2007). Strongly correlated two-photon transport in a one-dimensional waveguide coupled to a two-level system. *Physical Review Letters, 98*, 153003.
34. Shi, T., Fan, S., & Sun, C. P. (2011). Two-photon transport in a waveguide coupled to a cavity in a two-level system. *Physical Review A, 84*, 063803.
35. Tame, M. S., McEnery, K. R., Özdemir, Ş. K., Lee, J., Maier, S. A., & Kim, M. S. (2013). Quantum plasmonics. *Nature Physics.*
36. Wang, Y., Zhang, Y., Zhang, Q., Zou, B., & Schwingenschlogl, U. (2016). Dynamics of single photon transport in a one-dimensional waveguide two-point coupled with a Jaynes-cummings system. *Scientific Reports.*
37. Xiao, Y., Meng, C., Pan Wang, Y., Ye, H. Y., Wang, S., Fuxing, G., & Dai, L. (2011). Limin Tong.single-nanowire single-mode laser. *Nano Letters.*

Part V
Sustainable Infrastructure

Chapter 13
Intransitive Technology of the Plant's Transpiration Mechanism to Meet the Global Water Perplexity

Introduction

Plants take carbon dioxide from the atmosphere by the process of photosynthesis to release oxygen into the atmosphere to save the planet. Hence, plants cannot develop without water because it is essential for biological metabolic reactions need water [2, 9, 14]. Consequently, groundwater is taken up by plants via their roots and, via osmosis, carried to their leaves by means of the xylem [7, 13, 18]. Interestingly, plants only use around 0.3–5% of the water they take in for metabolic processes; the rest is discharged into the air through transpiration, which occurs through stomatal cells [11, 15, 17]. The greatest loss of groundwater is occurring as a result of transpiration, which is also adding to climate change because water vapor is a major greenhouse gas. Groundwater and transpiration, more especially terrestrial water flows, have been the subject of recent studies in rural regions. The water models have demonstrated that streamflow is decreasing due to plant transpiration [3, 5, 19]. There has been no investigation into a technique to collect this transpiration water to meet the world's water requirement, despite the fact that these findings are intriguing. This study presents a method to prevent water loss through transpiration by capturing water molecule instead of letting it escape into the air and then converting it into drinkable water. The placement of a plastic tank capable of producing static electricity next to the plants has been proposed for every rural residence to collect all of the water vapor that is drawn to static electricity [4, 6, 10]. To satisfy the daily water requirement, water vapor is drawn to a tank for on-site treatment by virtue of the attraction between the positive charge of static electricity and the negative charge of water. This occurs because the electrons that join the static electric field have a positive charge, and water vapor itself possesses both positive and negative charges. On the other hand, one side of each water molecule is negatively charged.

© The Author(s), under exclusive license to Springer Nature Switzerland AG 2025

M. F. Hossain, *Sustainable Global Development in Environment, Energy, Building, and Infrastructure*, https://doi.org/10.1007/978-3-031-84429-4_13

Materials and Methods

Static Electric Force Formation

In this research, an advanced engineering technology has been suggested to generate the Hossain Static Electric Force (**HSEF** = $\mathfrak{h}$) to collect the water vapor produced into the air by plants' transpiration process during the day. In this concept, the water molecule is tugged down by the force of static electricity from the tank installed at the bottom of the tree [1, 8]. Then, respecting gauge field asymmetry of the plastic tank and its Goldstone scalar quantity, the local abelian symmetries have been computed using MATLAB to produce Hossain Static Electric Force (HSEF) in the plastic tank [12, 19]. Once the locally asymmetric particle of the plastic tank is initiated due to the electromagnetic force, the HSEF will start work within the local U(1) phase asymmetry by producing static electric force (q) with its soundings that can be expressed as

$$\mathfrak{h} = -\frac{1}{4} F_{\mu\nu}F^{\mu\nu} + D_{\mu}\Phi^{*} D^{\mu}\Phi - V(\Phi^{*}\Phi) \tag{13.1}$$

Where

$$\begin{aligned} D_{\mu}\Phi(x) &= \partial_{\mu}\Phi(x) + iqA_{\mu}(x)\Phi\,(x) \\ D_{\mu}\Phi^{*}(x) &= \partial_{\mu}\Phi^{*}(x) - iqA_{\mu}(x)\Phi^{*}(x) \end{aligned} \tag{13.2}$$

And

$$V(\Phi^{*}\Phi) = \frac{\lambda}{2}\,(\Phi^{*}\Phi)^{2} + m^{2}(\Phi^{*}\Phi) \tag{13.3}$$

Here, $\lambda > 0$ and $m^2 < 0$, therefore $\Phi = 0$ is a local minimum vector potential, while the maximum form of degenerated electromagnetic circle is $\Phi = \frac{v}{\sqrt{2}} * e^{i\theta}$,

$$v = \sqrt{\frac{-2m^{2}}{\lambda}}, \text{ any real } \theta \tag{13.4}$$

Therefore, the asymmetry of the static electric field generation by the vector field, resulting from the spontaneous development of the U(1) asymmetry of the electromagnetic field of the plastic tank, will form a static electrical force that can be expressed

$$\Phi\,(x) = \frac{1}{\sqrt{2}}\,\Phi_{r}(x) * e^{i\Theta(x)}, \text{real } \Phi_{r}(x) > 0, \text{real } \Phi(x) \tag{13.5}$$

This force is identically real force Φ (x) with its sounding diameter, so, it can be considered the real fields' statice electricity force of the vector potential and denoted as

$$V(\phi) = \frac{\lambda}{8}\left(\phi_r^2 - v^2\right)^2 + \text{const}, \tag{13.6}$$

or in the form of the radial field, it can be shifted as VEV, $\Phi_r(x) = v + \sigma(x)$,

$$\phi_r^2 - v^2 = (v + \sigma)^2 - v^2 = 2v\sigma + \sigma^2 \tag{13.7}$$

$$V = \frac{\lambda}{8}\left(2v\sigma - \sigma^2\right)^2 = \frac{\lambda v^2}{2} * \sigma^2 + \frac{\lambda v}{2} * \sigma^3 + \frac{\lambda}{8} * \sigma^4 \tag{13.8}$$

Hence, the covariant derivative $D_\mu\phi$ *becomes*

$$D_\mu\phi = \frac{1}{\sqrt{2}}\left(\partial_\mu\left(\phi_r e^{i\Theta}\right) + iqA_\mu * \phi_r e^{i\Theta}\right) = \frac{e^{i\Theta}}{\sqrt{2}}\left(\partial_\mu\phi_r + \phi_r * i\partial_\mu\Theta + \phi_r * iqA_\mu\right) \tag{13.9}$$

$$\begin{aligned}\left|D_\mu\phi\right|^2 &= \frac{1}{2}\left|\partial_\mu\phi_r + \phi_r * i\partial_\mu\Theta + \phi_r * iqA_\mu\right|^2 = \frac{1}{2}\left(\partial_\mu\phi_r\right) + \frac{\phi_r^2}{2} * \left(\partial_\mu\Theta qA_\mu\right)^2 \\ &= \frac{1}{2}\left(\partial_\mu\sigma\right)^2 + \frac{(v+\sigma)^2}{2} * \left(\partial_\mu\Theta + qA_\mu\right)^2\end{aligned} \tag{13.10}$$

Altogether,

$$\mathfrak{h} = \frac{1}{2}\left(\partial_\mu\sigma\right)^2 - v(\sigma) - \frac{1}{4}F_{\mu\nu}F^{\mu\nu} + \frac{(v+\sigma)^2}{2} * \left(\partial_\mu\Theta + qA_\mu\right)^2 \tag{13.11}$$

To confirm that this HSEF's static electric field generation includes this static electric force in terms of their derivatives and the free particles' quadratic component as

$$\mathfrak{h}_{sef} = \frac{1}{2}\left(\partial_\mu\sigma\right)^2 - \frac{\lambda v^2}{2} * \sigma^2 - \frac{1}{4}F_{\mu\nu}F^{\mu\nu} + \frac{v^2}{2} * \left(qA_\mu + \partial_\mu\Theta\right)^2 \tag{13.12}$$

Here, the HSEF will surely form a functional mass scalar particle, which will use electromagnetic force to produce a large static electric force inside the plastic tank system (Fig. 13.1).

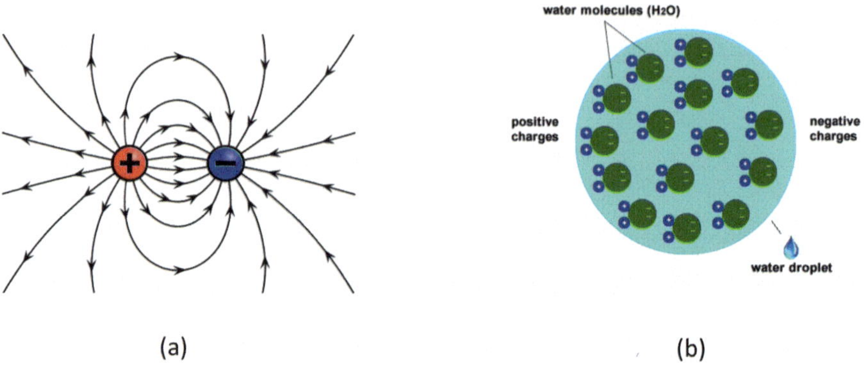

Fig. 13.1 (**a**) The process of static electricity generation. (**b**) The primary functional derivative to tug down the water in the plastic tank

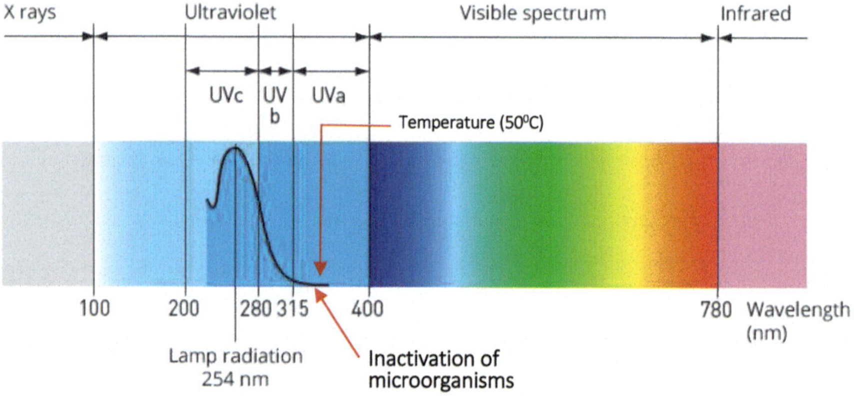

Fig. 13.2 When the water reaches 50 °C under UV light of 320 nm, all microorganisms get disinfected immediately

In Site Water Treatment

The accumulated water in the tank due to static electric force does not need to be treated with sedimentation, coagulation, or chlorination because it is only a liquid state of vapor. Through a simple combination of filtration and UV technology application, this water will be purified to meet the requirements of the US National Primary Drinking Water Standard code [12, 16]. This is the simplest way to purify water through container that is filled with water and left in direct sunlight for an hour. Once the water reaches 50 °C, it will react with 320 nm UV radiation of solar energy to speed up the inactivation process of all microbiological biochemical process inside the water if any and the water can be further undergone for filtration process to be considered the fully treated water (*see* Fig. 13.2).

Results and Discussion

Electrostatic Force Analysis

To confirm generation of static electric force sounding the tank to drag down the water inside it, the dynamics of electromagnetic force formed by the integrating HSEF has been clarified computationally. Subsequently, a variable gauge particle respecting the local U(1) gauge function has been calculated to get a better analysis of the covariant derivative that applies a unique transformation rule to the scalar field (∂_μ) due to the static electricity force [6, 15]:

$$\begin{aligned} \partial_\mu \rightarrow D_\mu = \partial_\mu = ieA_\mu \quad & [\text{covariant derivatives}] \\ A'_\mu = A_\mu + \frac{1}{e}\,\partial_\mu \alpha \quad & [A_\mu \text{ derivatives}] \end{aligned} \tag{13.13}$$

Here, the local U(1) gauge variant HSEF is a standard scalar field that can be written as

$$\mathfrak{h} = (D^\mu)^\dagger\,(D_\mu \varnothing) - \frac{1}{4}\,F_{\mu\nu}F^{\mu\nu} - V(\varnothing) \tag{13.14}$$

The term $\frac{1}{4}\,F_{\mu\nu}F^{\mu\nu}$ is the dynamic form for the gauge field and $V(\varnothing)$ represents the scalar form of HSEF that can be written as $V(\varnothing^*\varnothing) = \mu^2(\varnothing^*\varnothing) + \lambda\,(\varnothing^*\varnothing)^2$.

Hence, the HSEF electric field will be ready to pull down the water by introducing the active static electricity force in the tank in terms of the $\mathfrak{h}$ related to the scalar field [3, 8].

Thus, functional variable of the scalar field of this static electricity force confirms the resultant force of $\overline{\phi}\,[s_0]$ that is active for tugging down the water from the transpiration vapor of the tree as.

$$\begin{aligned} \overline{\phi}[s_0] = {} & 2s_0(\ln 4s_0 - 2) + \ln 4s_0(\ln 4s_0 - 2) - \frac{(\pi^2 - 9)}{3} + s_0^{-1}\left(ln\,4s_0 + \frac{9}{8}\right) \\ & + \cdots (s_0 \gg 1); \end{aligned} \tag{13.15}$$

$$\overline{\phi}[s_0] = \left(\frac{2}{3}\right)(S_0 - 1)^{\frac{3}{2}} + \left(\frac{5}{3}\right)(S_0 - 1)^{\frac{5}{2}} - \left(\frac{1507}{420}\right)(S_0 - 1)^{\frac{7}{2}} + \cdots (s_0 - 1 \ll 1) \tag{13.16}$$

Here, the function is defined as $1 < s0 < 10$, with the natural logarithmic value of $\overline{\phi}[s_0]$ included for greater s0, and thus, this value is being used to confirm that the HSEF successfully pulled down 100% water vapor into the tank.

In a statistic, average 100 gallons of water is needed each day for person [7, 16]. Based on the calculation (100 gallons per day per person × 4 persons × 365 days), a modest family of four will require a total of 146,000 gallons of water annually. Given that a typical oak tree can transpire 40,000 gallons (151,000 L) of water annually, the HSEF described earlier will only need four normal oak trees to meet a family of four's water needs.

Conclusions

The world is currently facing a serious water crisis due to the lowering of the groundwater strata by 30 feet over the last 30 years. It would be a fresh, integrated, and inventive field in science to use electrostatic force formed by the installed plastic tank to catch transpiration water molecule and then purify it on-site with a simple technology of UV application with the frequency of 320 nm and then application of standard filtration process. Simply put, this rerouting mechanism of the transpiration system will in fact be the solution to address these issues to facilitate the transformation and conversion of this water molecule into clean water to satisfy the demand for potable water on a worldwide scale.

Acknowledgments Regarding the publication of this chapter by an appropriate publishing company, the author affirms that there is no potential for a conflict of interest.

References

1. Andreas, R. (2012). Strongly correlated photons on a chip. *Nature Photonics, 6*, 93–96.
2. Douglas, S., Habibian, H., et al. (2015). Quantum many-body models with cold atoms coupled to photonic crystals. *Nature Photonics, 9*, 326–331.
3. Hossain, F. (2016). Theory of global cooling. *Energy, Sustainability and Society, 7*, 6–24.
4. Hossain, M. (2016). Solar energy integration into advanced building design for meeting energy demand and environment problem. *International Journal of Energy Research, 40*, 1293–1300.
5. Hossain, M. F. (2017). Green science: Independent building technology to mitigate energy, environment, and climate change. *Renewable and Sustainable Energy Reviews, 73*, 695–705.
6. Jaivime, E., Scott, J., & McDonnell, J. J. (2015). Global separation of plant transpiration from groundwater and streamflow. *Nature, 525*, 91–94.
7. Josette, M., & Scott, R. (2005). The ERECTA gene regulates plant transpiration efficiency in Arabidopsis. *Nature, 436*, 866–870.
8. Langer, L., Poltavtsev, S., & Bayer, M. (2014). Access to long-term optical memories using photon echoes retrieved from semiconductor spins. *Nature Photonics, 8*, 851–857.
9. Leijing, Y., Sheng, W., Qingsheng, Z., Zhiyong, Z., Tian, P., & Yan, L. (2011). Efficient photovoltage multiplication in carbon nanotubes. *Nature Photonics, 8*, 672–676.
10. Li, Q., & Xu, D. (2013). Recoil effects of a motional scatterer on single-photon scattering in one dimension. *Scientific Reports, 8*, 3144.
11. Maxwell, R. M., & Condon, L. E. (2015). Connections between groundwater flow and transpiration partitioning. *Science, 353*, 377–380.

12. Pregnolato, T., Lee, E., Song, J., Stobbe, D., & Lodahl, P. (2015). Single-photon non-linear optics with a quantum dot in a waveguide. *Nature Communications, 6*, 8655.
13. Scott, J., & Zachary, D. (2013). Terrestrial water fluxes dominated by transpiration. *Nature, 496*, 347–350.
14. Soto, W., Klein, S., et al. (2006). Improvement and validation of a model for photovoltaic array performance. *Solar Energy, 80*, 78–88.
15. Tame, M., McEnery, S., et al. (2013). Quantum plasmonics. *Nature Physics, 9*, 329–340.
16. Wheeler, T. D., & Stroock, A. D. (2008). The transpiration of water at negative pressures in a synthetic tree. *Nature, 455*, 208–212.
17. Yan, W., & Heng, F. (2014). Single-photon quantum router with multiple output ports. *Scientific Reports, 4*, 4820.
18. Yuwen, W., Yongyou, Z., Qingyun, Z., Bingsuo, Z., & Udo, S. (2016). Dynamics of single photon transport in a one-dimensional waveguide two- point coupled with a Jaynes-cummings system. *Scientific Reports, 6*, 33867.
19. Zhu, Y., Xiaoyong, H., Hong, Y., & Qihuang, G. (2014). On-chip plasmon-induced transparency based on plasmonic coupled nanocavities. *Scientific Reports, 4*, 3752.

Chapter 14
Simulation of the Environmental Force for Development of Global Infrastructure System

Introduction

The geomagnetic field, commonly referred to as earth's magnetic field, emanates from the earth's interior and it reaches into outer space [20, 34]. This magnetic field forms magnetosphere resulting from the convection currents of the molten iron and nickel alloy in earth's core [8, 11]. Interestingly, this magnetosphere is delineated by the reach of earth's magnetic field, extending above the ionosphere for several tens of thousands of kilometers into space [18, 32]. Hence, the North geomagnetic pole of the earth signifies the South Pole, while the South geomagnetic pole represents the North pole of earth's magnetic field, as opposite magnetic poles attract each other. The intensity of earth's magnetic field at its surface varies from 25 to 65 μT that is characterized by a magnetic dipole inclined approximately 11.5° relative to earth's rotational axis, analogous to a colossal bar magnet positioned at that angle through the earth's center [7, 9]. To utilize this tremendous natural magnetic power, in this study, earth's electromagnetic field has been computed to develop a model for the invisible infrastructure throughout the world above the earth's surface to run energy state travel machine on it.

Simply in this work, a bridge of the earth's electromagnetic field, spherical harmonics of the earth's electromagnetic field, and Higgs–Boson's electromagnetism-assisted air transport have been modeled to elucidate how the electromagnetic field can be decoded into invisible infrastructure. Subsequently, how the air transport can run over this infrastructure system to travel anywhere on the planet.

© The Author(s), under exclusive license to Springer Nature Switzerland AG 2025

M. F. Hossain, *Sustainable Global Development in Environment, Energy, Building, and Infrastructure*, https://doi.org/10.1007/978-3-031-84429-4_14

Methods and Simulation

Earth's Electromagnetic Force

The electromagnetic field of the earth is being modeled by a three-dimensional vector that includes X (North), Y (East), and Z (Down) coordinates to computationally elucidate the magnetic force, taking into account the entire plane (*see* Fig. 14.1).

Subsequently, in the control mechanism of this electromagnetic field considering the earth's sphericity at its positive x-direction of vector, a mathematical modeling is being performed and expressed by following equation as

$$E\,(x,t) = E_{\max}\cos(kx - \omega t + \phi) \tag{14.1}$$

Here, the scalar of the electromagnetic field is

$$B\,(x,t) = B_{\max}\cos(kx - \omega t + \phi) \tag{14.2}$$

Hence, by implementation of the Maxwell equation for electromagnetism, the equation can be rewritten as

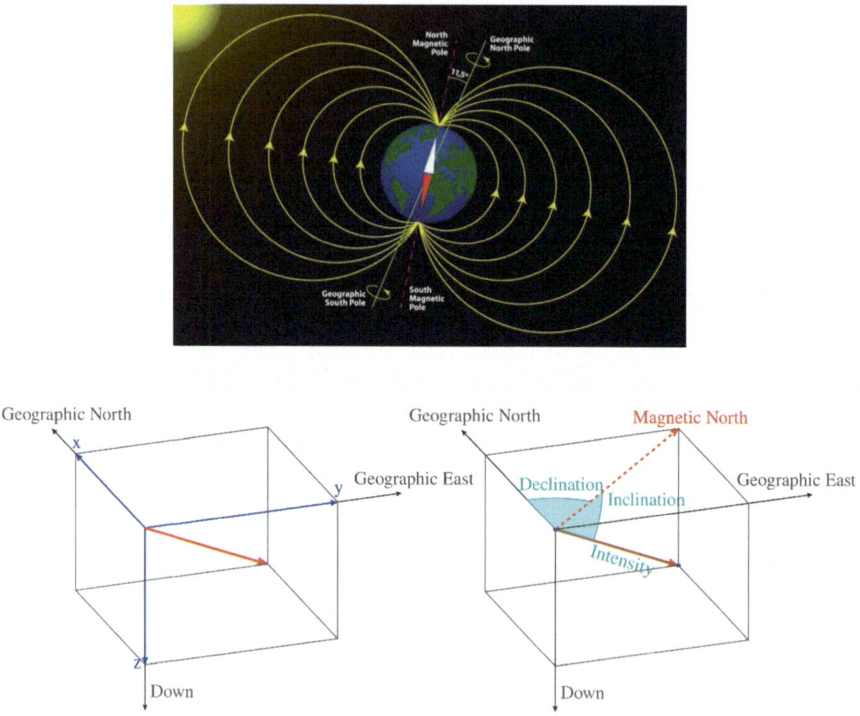

Fig. 14.1 The concept of the coordinate systems of the earth's electromagnetic field

$$\nabla.E = \frac{\rho}{\varepsilon_0} \tag{14.3}$$

$$\nabla.B = 0 \tag{14.4}$$

$$\nabla \times E = -\frac{\partial B}{\partial t} \tag{14.5}$$

$$\nabla \times B = \mu_0\left(J - \varepsilon_0\frac{\partial E}{\partial t}\right) \tag{14.6}$$

where E is the electric field, ρ is the charge density, ε_0 is a constant of the permeability, B is the magnetic field, μ_0 is another contact related to the permeability, and J is the electric current density. Thus, in a plane ($\rho = 0$, $J = 0$), the following electromagnetic field can be rewritten as

$$E = <0, A\cos\left\{\frac{2\pi}{\tau}(x - ct)\right\}, 0> \tag{14.7}$$

$$B = <0, 0, \frac{A}{c}\cos\left\{\frac{2\pi}{\tau}(x - ct)\right\}, 0> \tag{14.8}$$

where A, τ, and c are constant (amplitude, wavelength, and speed of light). As c is related to μ_0 and ε_0 is related to $\mu_0\varepsilon_0 = c^{-2}$, the plane derived from the earth's electromagnetic force is being reformulated here as R_E, taking into account the distance R_O and the average $(a.u.)$ and the force constant S_O [2, 14, 26]. Consequently, the electromagnetic field on earth plane is being computed as

$$Q = \begin{cases} S_o\frac{R_0^2}{R_E^2}\cos(\theta) & \cos(\theta) > 0 \\ 0 & \cos(\theta) \leq 0 \end{cases} \tag{14.9}$$

Here, Q is the electromagnetic field in coordination with its vector considering a mean time of $h = \pi$ to $h = -\pi$, and thus, the equation can be formulated as

$$Q^{\text{time}} = \frac{1}{2\pi}\int_{\pi}^{-\pi} Qdh \tag{14.10}$$

Hence, h_O is the earth plane and Q is the functional integer, the electromagnetic field on the earth plane could be solved as

$$\rho\sin(\phi)\sin(\delta) + \cos(\phi)\cos(\delta)\cos(h_0) = 0 \tag{14.11}$$

or

$$\rho\cos(h_0) = -\tan(\phi)\tan(\delta) \tag{14.12}$$

As tan(φ) tan(δ) > 1, the electromagnetic field considering the meantime $h = \pi$ and $ho = \pi$ in a $Q^{\text{time}} = 0$, the earth plane $\frac{R_0^2}{R_E^2}$ could be denoted as

$$\int_{\pi}^{-\pi} Qdh = \int_{h_0}^{-h_0} Qdh = S_0 \frac{R_0^2}{R_E^2} \int_{h_0}^{-h_0} \cos(\theta) dh$$

$$= S_0 \frac{R_0^2}{R_E^2} [h\sin(\phi)\sin(\delta) + \cos(\phi)\cos(\delta)\sin(h)] \begin{matrix} h = -h_0 \\ h = h_0 \end{matrix} \tag{14.13}$$

Therefore

$$Q^{\text{time}} = \frac{S_0}{\pi} \frac{R_0^2}{R_E^2} \left[h_0 \sin(\phi)\sin(\delta) + \cos(\phi)\cos(\delta)\sin\left(h_{0)}\right]\right. \tag{14.14}$$

Hence, θ represents here as the sphere of electromagnetic field of the vector of the plane, thus, θ = 0 is related to its vernal equinox of δ at its plane that can be expressed as

$$\delta = \varepsilon \sin(\theta) \tag{14.15}$$

Here, ε represents the longitude of plane and ω represents the equinox. So, the equation could be addressed as

$$R_E = \frac{R_0}{1 + e\cos(\theta - \omega)} \tag{14.16}$$

or

$$\frac{R_0}{R_E} = 1 + e\cos(\theta - \omega) \tag{14.17}$$

Considering the aforementioned ω, ε, s_o, and e from *a.u.* computation, the electromagnetic field is being computed respecting any latitude φ and longitude θ of the earth plane [4, 30, 35]. As θ = 0° is denoting here the equinox and θ = ∞ is denoting the plane of the earth, the equation can be rewritten for electromagnetic field as

$$M_e = S_0 \left(1 + \cos\left(2\pi \frac{n}{x}\right)\right) \tag{14.18}$$

Naturally, here S_o is the force constant, n is the unit of electromagnetic field, x represents the electromagnetism, and π is the plane of the electromagnetic field that corresponds to the electromagnetic force (I_p) of the earth that paves the formation of the invisible infrastructure (M_e).

Invisible Infrastructure

Subsequently, the analysis of global variables of magnetic field of the earth has been computed to accommodate the creation of invisible infrastructure by employing a series of spherical harmonics of its electromagnetic field [6, 28, 36]. As spherical harmonics are the functions that oscillate across the surface of the sphere, considering the latitude and longitude of the earth, and are influenced by any vector fields corresponding to the Cartesian coordinates X, Y, and Z, each component of the singular scalar function could be considered as the magnetic potential [1, 24, 29]. As the spherical harmonics is the hidden particle, it has been clarified by the Laplacian of scalar quantities (V) with respect to its tangential (T). Subsequently, the novel U (1) gauge symmetry associated with spherical harmonics can be represented as an invisible infrastructure by articulating the flow equation:

$$\overrightarrow{De} = \left[\frac{Edp}{c}, \ \overrightarrow{Pd}\ \right] + \overrightarrow{Xef}\left[\frac{Eef}{c}, \overrightarrow{Pef}\ \right] \tag{14.19}$$

As spherical harmonics is dynamics, $\Delta P = 0$, the sum of the spherical harmonics must be equal to the sum of the final momentum that can be addressed as

$$\overrightarrow{De} + \overrightarrow{Xei} = \overrightarrow{Xp} + \overrightarrow{Xef} \tag{14.20}$$

Multiplying the spherical harmonics of the scale considering with its cross terms, it can be addressed as

$$2\,\overrightarrow{Xd} \cdot \overrightarrow{Xei} = 2 * \left[\frac{Ed}{c}, \overrightarrow{Pd}\right] \cdot \left[m_e c, \vec{0}\,\right] = 2Edm_e \tag{14.21}$$

Consequently, integrating all the momentums of the electromagnetic field, the generation of spherical harmonics will form invisible infrastructure globally that can be written as

$$m_d^2c^2 - m_e^2c^2 + -2m_eE_p - 2\left(\frac{E_dE_p}{c^2} - \frac{P_dE_p\cos(\theta)}{c}\right) + 2E_dm_e = m_e^2c^2 \tag{14.22}$$

As the formulation of the invisible infrastructure is correlated to the scalar momentum of its electromagnetic force of the entire world, the spherical harmonics will pave the vehicle to run over the invisible infrastructure (see Fig. 14.2).

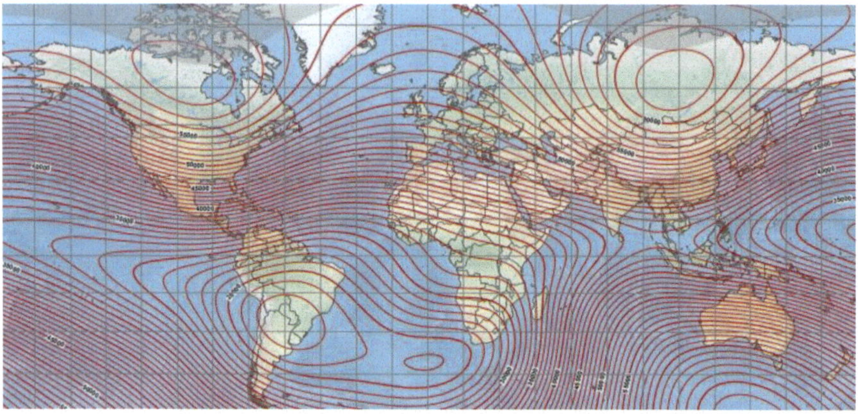

(a)

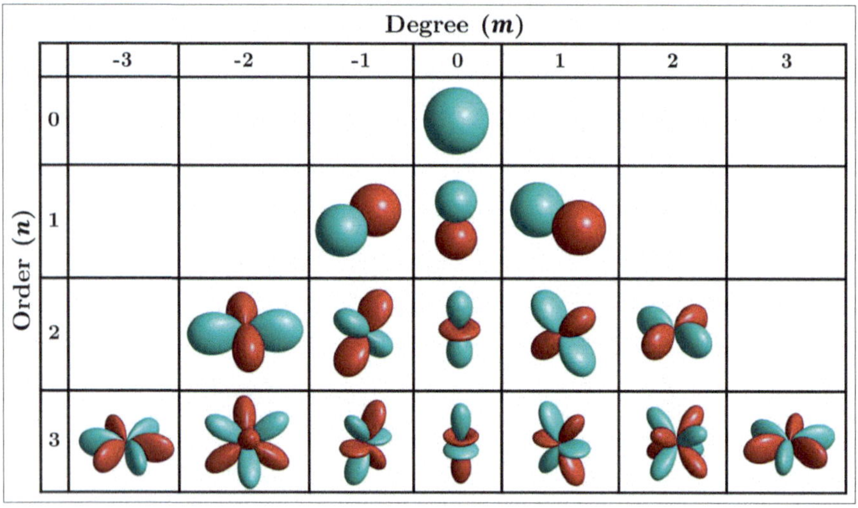

(b)

Fig. 14.2 The conceptual development model of the invisible infrastructure respecting (**a**) word magnetic field and (**b**) spherical harmonics of the earth's electromagnetic force

Air Dynamics Transport

Subsequently, air dynamics is being modeled as the transport by employing Higgs–Boson's electromagnetism in the local symmetry of U(1) of the vectorial momentum of electromagnetic force to travel on the invisible infrastructure [3, 23, 27]. Thus, the air dynamic force generation is being considered as the energy state travel machine (see Fig. 14.3).

Fig. 14.3 The concept of air dynamics transport to be run on the invisible infrastructure by the force of vectorial momentum of the Higgs–Boson's electromagnetism

Hence, the air dynamic energy state travel machine that is air dynamics can be further addressed by implementing the Klein and Gordon (*K–D*) equation as

$$\left((-i\hbar\nabla)^2c^2 + m^2c^4\right)\psi = \left(i\hbar\frac{\partial}{\partial t}\right)^2\psi \tag{14.23}$$

which simplifies to

$$-\hbar^2c^2\nabla^2\psi + m^2c^4\psi = \hbar^2\frac{\partial^2}{\partial t^2} \tag{14.24}$$

Rearranging terms yields

$$c\psi + \frac{\partial V}{\partial \psi} = 0 \tag{14.25}$$

Subsequently, the proof *K–D* equation conceding an algebraic term can be rewritten as follows:

$$\left(c + \mu^2\right)\varphi(x) = 0 \tag{14.26}$$

Then, multiplying by the scalar quantities $\varphi^*(x)$ and $\varphi(x)$, it can be rewritten as

$$\varphi^*(c+\mu^2)\varphi = 0 \tag{14.27}$$

$$\varphi(c+\mu^2)\varphi^* = 0 \tag{14.28}$$

Subtracting the former from the latter, it can be obtained

$$\varphi^* c\varphi - \varphi c\varphi^* = 0 \tag{14.29}$$

$$\varphi^*\partial_\mu\partial^\mu\varphi - \varphi\partial_\mu\partial^\mu\varphi^* = 0 \tag{14.30}$$

Here, the formation of the scalar variable to the *K–D* field, the vectorial air dynamic can be addressed as follows:

$$T^{\mu\nu} = \frac{\hbar^2}{m}\left(\eta^{\mu\alpha}\eta^{\nu\beta} + \eta^{\mu\beta}\eta^{\nu\alpha} - \eta^{\mu\nu}\eta^{\alpha\beta}\right)\partial_\alpha\overline{\psi}\partial_\beta\psi - \eta^{\mu\nu}mc^2\overline{\psi}\psi \tag{14.31}$$

Hence, the force component T^{00} is related to the energy state travel machine that is air dynamic $\psi(x,t) = \varnothing(x,t)e^{-\frac{i}{\hbar}mc^2t}$ where $\varnothing(x,t) = u_E(x)e^{-\frac{i}{\hbar}Et}$. Therefore, the air dynamic energy state travel machine (T_m) on the invisible infrastructure $E' = E - mc^2 = \sqrt{m^2c^4 + c^2p^2} - mc^2 \approx \frac{p^2}{2m}$, $E' \ll mc^2$ should have a nonrelativistic limit $v\sim p < < c$, $\left|i\hbar\frac{\partial\varnothing}{\partial t}\right| = E'\varnothing \ll mc^2\varnothing$; thus, the derivative of the air dynamics ψ can be written as

$$\begin{aligned} \frac{\partial\psi}{\partial t} &= \left(-i\frac{mc^2}{\hbar}\varnothing + \frac{\partial\varnothing}{\partial t}\right)e^{-\frac{i}{\hbar}mc^2t} \approx -i\frac{mc^2}{\hbar}\varnothing e^{-\frac{i}{\hbar}mc^2t} \\ \frac{\partial^2\psi}{\partial t^2} &\approx \left(i\frac{2mc^2}{\hbar}\frac{\partial\varnothing}{\partial t} + \left(\frac{mc^2}{\hbar}\right)^2\varnothing\right)e^{-\frac{i}{\hbar}mc^2t} \end{aligned} \tag{14.32}$$

Substituting it with the free *K–D* equation, $c^{-2}\partial_t^2\psi = \nabla^2\psi - m^2\psi$, energy state travel machine $\frac{1}{c^2}\left(i\frac{2mc^2}{\hbar}\frac{\partial\varnothing}{\partial t} + \left(\frac{mc^2}{\hbar}\right)^2\varnothing\right)e^{-\frac{i}{\hbar}mc^2t} \approx \left(\nabla^2 - \left(\frac{mc^2}{\hbar}\right)^2\right)\varnothing e^{-\frac{i}{\hbar}mc^2t}$ can be simplified as

$$-i\hbar\frac{\partial\varnothing}{\partial t} = -\frac{\hbar^2}{2m}\nabla^2\varnothing \tag{14.33}$$

Subsequently, the quantum field of the Klein–Gordon equation here confirms that the air dynamic energy state travel machine is functionable on the invisible structure of $\varphi \rightarrow \varphi' = \exp(i\theta)\varphi$ where $\theta(t,x)$ is a local scalar angle considering its functional dynamics of $\exp(i\theta) = \cos\theta + i\sin\theta$, where derivative ∂_μ and gauge-covariant variable of $D_\mu = \partial_\mu - ieA_\mu$ to confirm that energy state travel machine (T_m) can run over the invisible infrastructure as

$$D_\mu D^\mu \varphi = -(\partial_t - ieA_0)^2 \varphi + (\partial_i - ieA_i)^2 \varphi = m^2 \varphi \tag{14.34}$$

Results and Discussion

Earth's Electromagnetic Force

The mechanism of the three-dimensional vector comprising X (North), Y (East), and Z (Down) coordinates has been computed mathematically to clarify the electromagnetic field of the earth by the implementing process of current loops generation magnetic fields, changing magnetic field, and the electromagnetic fields exert force [16, 22, 23]. Thus, electromagnetic field of the earth is being denoted as follows:

$$\frac{\partial B}{\partial t} = \eta \nabla^2 B + (u + B) \tag{14.35}$$

where u is the velocity of the magnetic force, B is the magnetic field, and $\eta = \frac{1}{\sigma\mu}$is the magnetic intensity, and σ is the electrical conductivity. The term $\frac{\partial B}{\partial t}$here is the partial derivative of the electromagnetic field related to time t; and the vector ∇^2.

As this earth's magnetic field is formed by electric currents in the conductive iron alloys of its core, a particle with charge q moving in an electromagnetic field can be described as

$$L = \frac{mv^2}{2} + \frac{q}{c} v.A(r,t) - q\varnothing(r,t) \tag{14.36}$$

Here, the electromagnetic field is related to the dynamic force of its electromatic field, thus, the equation can be addressed as

$$E = -\frac{1}{c}\frac{\partial A}{\partial t} - \nabla\varnothing, B = \nabla \text{x A} \tag{14.37}$$

and

$$\text{v x } (\nabla \text{x A}) = \nabla(\text{v.A}) - (\text{v.}\nabla)\text{A} \tag{14.38}$$

Consequently, the electromagnetic field of the earth can be confirmed by employing Maxwell's theory of electromagnetic field of $\varrho_{PC}(\omega) \propto \frac{1}{\sqrt{\omega - \omega_e}} \Theta(\omega - \omega_e)$, where $\Theta(\omega - \omega_e)$ and ω_e are the intensity of the electric field [5, 22, 33]. Then, the intensity of the electromagnetic field is being clarified to confirm the qualitative form of electromagnetic field by employing the Weisskopf–Winger module of $\varrho_{PC}(\omega) \propto \frac{1}{\sqrt{\omega - \omega_e}} (\omega - \omega_e)$. Here, the electromagnetic field is a perfect algorithmic, which is $\varrho_{PC}(\omega) \propto -[\text{ln}|(\omega - \omega_0)/\omega_0| - 1]\Theta(\omega - \omega_e)$, thus, the primary function of determining the intensity of the electromagnetic field of earth can be written as follows:

$$J(\omega) = \varrho(\omega)|V(\omega)|^2. \tag{14.39}$$

As the electromagnetic field of ω_c and the intensity of the electromagnetic field $\langle a(t)\rangle = u(t, t_0)\langle a(t_0)\rangle$, the function of the electromagnetic field $u(t, t_0)$ can be described as the combined electromagnetic field, $u(t, t_0)$, by addressing the following equation:

$$u(t, t_0) = \frac{1}{1 - \Sigma'(\omega_b)} e^{-i\omega(t - t_0)} + \int_{\omega_e}^{\infty} d\omega \frac{J(\omega) e^{-i\omega(t - t_0)}}{[\omega - \omega_c - \Delta(\omega)]^2 + \pi^2 J^2(\omega)}, \tag{14.40}$$

where $\Sigma'(\omega_b) = [\partial\Sigma(\omega)/\partial\omega]_{\omega = \omega_b}$ and $\Sigma(\omega)$ is the electromagnetic field, and the question can be simplified as

$$\Sigma(\omega) = \int_{\omega_e}^{\infty} d\omega' \frac{J(\omega')}{\omega - \omega'} \tag{14.41}$$

Here, the frequency ω_b represents the intensity of the electromagnetic field where $(0 < \omega_b < \omega_e)$ that is the electromagnetic frequency mode, thus, it has been calculated as $\omega_b - \omega_c - \Delta(\omega_b) = 0$, where $\lesssim\Delta(\omega) = \mathcal{P}\left[\int d\omega' \frac{J(\omega')}{\omega - \omega'}\right]$ is a primary integral of the electromagnetic field [8, 31, 36]. Naturally, it can be explained that electromagnetic field (eV) on the earth is Maxwell's electric field n_0, *i. e.* , $\rho(t_0) = |$ $n_0\rangle\langle n_0|$, which can be addressed considering the real-time electromagnetic field of earth's radial (r) at a time t as

$$\rho(t) = \sum_{n=0}^{\infty} \mathcal{P}_n^{(n_0)}(t)|n_0\rangle\langle n_0| \tag{14.42}$$

and

$$\mathcal{P}_n^{(n_0)}(t) = \frac{[v(t,t)]^n}{[1 + v(t,t)]^{n+1}} [1 - \Omega(t)]^{n_0} \text{ x } \sum_{k=0}^{\min\{n_0, n\}} \binom{n_0}{k} \binom{n}{k} \left[\frac{1}{v(t,t)} \frac{\Omega(t)}{1 - \Omega(t)}\right]^k \tag{14.43}$$

As $\Omega(t) = \frac{|u(t, t_0)|^2}{1 + v(t,t)}$ is an invisible force, this result suggested that this electromagnetic field of the earth $\mathcal{P}_n^{(n_0)}(t)$ of $|n_0\rangle$ is the confirmation state of the electromagnetic field of $\mathcal{P}_n^{(n_0)}(t)$ which could be utilized to form invisible infrastructure.

Invisible Infrastructure

To make earth's electromagnetic field suitable for the formation of invisible infrastructure, a collection of spherical harmonics of this field has been computed. Hence, spherical harmonics can be used to measure all dimensions of the forces considering the latitude and longitude and the Cartesian components of X, Y, Z [10, 13, 15]. Thus, the scalar field of electromagnetic field considering its tangential (T) height from the earth surface can be written by Maxwell's equation as

$$\frac{\partial^2 E}{\partial x^2} = \mu\varepsilon \frac{\partial^2 E}{\partial t^2} \tag{14.44}$$

$$\frac{\partial^2 B}{\partial x^2} = \mu\varepsilon \frac{\partial^2 B}{\partial t^2} \tag{14.45}$$

Simplified these differential equations, the spherical harmonics of electromagnetic force can be written as

$$E = E_{\max} \cos(kx - \omega t); B = B_{\max} \cos(kx - \omega t) \tag{14.46}$$

In spherical harmonics, the local symmetry of electromagnetic field has been modeled considering local Abelian symmetries by implementing Higgs–Boson electromagnetic field [12, 15, 21]. Therefore, the local symmetries of spherical harmonics of the electromagnetic field will generate a scalar field $\Phi(x)$ to generate charge q on the electromagnetic field $A^{\mu}(x)$ in order to pave an invisible infrastructure that can be written as

$$\mathcal{L} = -\frac{1}{4} F_{\mu\nu} F^{\mu\nu} + D_{\mu}\Phi^* D^{\mu}\Phi - V(\Phi^*\Phi), \tag{14.47}$$

where

$$\begin{aligned} D_{\mu}\Phi(x) &= \partial_{\mu}\Phi(\mathrm{x}) + \mathrm{iqA_{\mu}(x)}\Phi\,(\mathrm{x}) \\ D_{\mu}\Phi^*(x) &= \partial_{\mu}\Phi^*(\mathrm{x}) - \mathrm{iqA_{\mu}(x)}\Phi^*(\mathrm{x}), \end{aligned} \tag{14.48}$$

and

$$V(\Phi^*\Phi) = \frac{\lambda}{2}\,(\Phi^*\Phi)^2 + m^2(\Phi^*\Phi). \tag{14.49}$$

As $\lambda > 0$; $m^2 < 0$, thus, $\Phi = 0$ which is a local vatable of the vector, the formation of spherical harmonics dynamics would be

$$v = \sqrt{\frac{-2m^2}{\lambda}} \text{ for any real } \theta. \tag{14.50}$$

Subsequently, the vector dynamics will generate a non-zero scalar field $\langle \Phi \rangle \neq 0$, which will address the local U(1) symmetries into spherical harmonics [17, 19, 27]. Therefore, for the development of invisible infrastructure, the spherical harmonics dynamics of Φ (x) can be expressed as

$$\Phi\ (\mathrm{x}) = \frac{1}{\sqrt{2}}\ \Phi_{\mathrm{r}}(\mathrm{x}) * \mathrm{e}^{\mathrm{i}\Theta(\mathrm{x})}, \text{real } \Phi_{\mathrm{r}}(\mathrm{x}) > 0, \text{real } \Phi(\mathrm{x}). \tag{14.51}$$

As the scalar field of invisible infrastructure is a dynamic spherical harmonic Φ $(\mathrm{x}) = 0$, it will address the principle of momentum of the force $\langle \Phi \rangle \neq 0$. So, the spherical harmonics of the invisible infrastructure can be addressed as

$$V(\phi) = \frac{\lambda}{8}\ \left(\phi_r^2 - v^2\right)^2 \tag{14.52}$$

Naturally, the momentum field can be replaced by employing the variable scalars, $\Phi_r(x) = v + \sigma(x)$, and expressed as

$$\phi_r^2 - v^2 = (v + \sigma)^2 - v^2 = 2v\sigma + \sigma^2 \tag{14.53}$$

$$V = \frac{\lambda}{8}\ \left(2v\sigma - \sigma^2\right)^2 = \frac{\lambda v^2}{2} * \sigma^2 + \frac{\lambda v}{2} * \sigma^3 + \frac{\lambda}{8} * \sigma^4. \tag{14.54}$$

Thus, the functional derivative $D_\mu \phi$ will become

$$D_\mu \phi = \frac{1}{\sqrt{2}}\ \left(\partial_\mu\left(\phi_r e^{i\Theta}\right) + iqA_\mu * \phi_r e^{i\Theta}\ \right) = \frac{e^{i\Theta}}{\sqrt{2}}\ \left(\partial_\mu \phi_r + \phi_r * i\partial_\mu \Theta + \phi_r * iqA_\mu\right) \tag{14.55}$$

$$\begin{aligned} \left|D_\mu \phi\right|^2 &= \frac{1}{2}\ \left|\partial_\mu \phi_r + \phi_r * i\partial_\mu \Theta + \phi_r * iqA_\mu\right|^2 \\ &= \frac{1}{2}\ \left(\partial_\mu \phi_r\right) + \frac{\phi_r^2}{2} * \left(\partial_\mu \Theta qA_\mu\right)^2 = \frac{1}{2}\ \left(\partial_\mu\ \sigma\right)^2 + \frac{(v + \sigma)^2}{2} * \left(\partial_\mu \Theta + qA_\mu\right)^2 \end{aligned} \tag{14.56}$$

Subsequently, the Lagrangian is then implemented here for the generation of invisible infrastructure through spherical harmonics to pave the travel machine to run on it, which is expressed as follows:

$$\mathcal{L} = \frac{1}{2}\left(\partial_\mu\ \sigma\right)^2 - v(\sigma) - \frac{1}{4} F_{\mu\upsilon} F^{\mu\upsilon} + \frac{(v + \sigma)^2}{2} * \left(\partial_\mu \Theta + qA_\mu\right)^2. \tag{14.57}$$

Air Dynamics Transport

Subsequently, air dynamics, in innovative transportation system has been confirmed computationally to run on the invisible infrastructure as a dynamic energy state travel machine. Necessarily, the spherical harmonics of electromagnetic field has been analyzed by employing Higgs–Boson's electromagnetism considering its local symmetry of U(1) analysis, in terms of gauge particles of invisible infrastructure $\varnothing' \to e^{i\alpha(x)}\varnothing$ to secure the air dynamics transports can run smoothly on the invisible infrastructure and can be expressed as.

$$\begin{aligned} \partial_\mu \to D_\mu = \partial_\mu = ieA_\mu \quad &[\text{covariant derivatives}] \\ A'_\mu = A_\mu + \frac{1}{e}\,\partial_\mu\alpha \quad &[A_\mu \text{ derivatives}] \end{aligned} \tag{14.58}$$

Here, the local U(1) gauge invariant Lagrangian is being employed as the perplex scalar field of the energy state travel machine and expressed by

$$\mathcal{L} = (D^\mu)^\dagger\,(D_\mu\varnothing) - \frac{1}{4}\,F_{\mu\nu}F^{\mu\nu} - V(\varnothing) \tag{14.59}$$

Hence, the term $\frac{1}{4}\,F_{\mu\nu}F^{\mu\nu}$ is the vectorial motion of energy state travel machine considering its vectorial gauge field $V(\varnothing)$; thus, the kinetic term of energy state travel machine can be rewritten as

$$V(\varnothing^*\varnothing) = \mu^2(\varnothing^*\varnothing) + \lambda\,(\varnothing^*\varnothing)^2 \tag{14.60}$$

As $\mu^2 < 0$, the energy state travel machine $\phi_0 = \frac{1}{\sqrt{2}}[(\upsilon + \eta) + i\xi]$ can run actively considering a kinetic mode of

$$\mathcal{L}(\eta,\xi) = (D^\mu\phi)^\dagger(D^\mu\phi) = (\partial^\mu + ieA^\mu)\phi^*\,(\partial_\mu - ieA_\mu)\,\phi \tag{14.61}$$

As $V(\eta,\xi) = \lambda\,\upsilon^2\eta^2$ is the final term of the kinetic mode of the energy state travel machine, it can be rewritten as:

$$\mathcal{L}(\eta,\xi) = \frac{1}{2}\,(\partial_\mu\eta)^2 - \lambda\upsilon^2\eta^2 + \frac{1}{2}\,(\partial_\mu\xi)^2 - \frac{1}{4}\,F_{\mu\nu}F^{\mu\nu} + \frac{1}{2}\,e^2\upsilon^2A_\mu^2 - e\upsilon A_\mu(\partial^\mu\xi) \tag{14.62}$$

Here, η represents the mass, ξ represents the massless, μ represents the mass for the quanta of the energy state travel machine, and A_μ is defined as the term $\partial_\mu\alpha$, as is the function of the energy state travel machine [17, 25, 31]. Naturally, A_μ and ϕ can be changed spontaneously, so the equation can be rewritten as a Lagrangian scalar to confirm the formation of the dynamic state travel machine (T_m) as

$$\mathcal{L}_{\text{scalar}} = (D^{\mu}\phi)^{\dagger}(D^{\mu}\phi) - V(\phi^{\dagger}\phi) = (\partial^{\mu} + ieA^{\mu})\frac{1}{\sqrt{2}}(v+h)\left(\partial_{\mu} - ieA_{\mu}\right)\frac{1}{\sqrt{2}}(v+h) - V(\phi^{\dagger}\phi) \tag{14.63}$$

$$= \frac{1}{2}\left(\partial_{\mu}h\right)^2 + \frac{1}{2}e^2A_{\mu}^2(v+h)^2 - \lambda v^2h^2 - \lambda vh^3 - \frac{1}{4}\lambda h^4 + \frac{1}{4}\lambda h^4 \tag{14.64}$$

Simply, this mathematical result confirms the representation of the energy state travel machine capable of traveling anywhere on the planet over the invisible infrastructure that in fact originated from the electromagnetic force of the earth.

Conclusions

The computation of electromagnetic force of the earth and its transformation mechanism has been modeled through a series of mathematical probes to develop invisible infrastructure to run energy state travel machine on it. The calculation revealed that the employment of spherical harmonics of the earth's electromagnetic force is the most promising ones for the development of invisible infrastructure field throughout the world. Subsequently, the use of the air dynamics as the energy state travel machine by applying Higgs–Boson's electromagnetism to the local U(1) symmetry of the invisible infrastructure shall represent the pinnacle of infrastructure systems ever thought, enabling rapid and safe travel across the planet in the near future.

Acknowledgments The author declares that this chapter has no reservation to be published by any appropriate publisher.

References

1. Abdelmalek, S., Barazane, L., Larabi, A., & Belmili, H., (2015). Contributions to diagnosis and fault tolerant control based on proportional integral observer: Application to a doubly-fed induction generator. In *2015 4th international conference on electrical engineering (ICEE)*.
2. Aldo, D., & Alfred, R. (1999). Design of an integrated electromagnetic levitation and guidance system for Swiss Metro, EPE'99, Lausanne, Swiss.
3. Astrom, K. J., & Hagglund, T. (1988). *Automatic tuning of PID controllers*. Instrument Society of America.
4. Barambones, O. (2013). Power output maximization for wave power generation plants using an adaptive sliding mode control. In *2013 International Conference on Renewable Energy Research and Applications (ICRERA)*, Madrid, pp. 178–183. https://doi.org/10.1109/ICRERA.2013.6749747
5. Boutotbat, M., Mokrani, L., & Machmoum, M. (2013). Control of a wind energy conversion system equipped by a DFIG for active power generation and power quality improvement. *Renewable Energy, 50*, 378–386.

6. Cai, Y., & Chen, S. S. (1997). Dynamic characteristics of magnetically levitated vehicle systems. *Applied Mechanics Reviews, 50*(11), 647–670.
7. Cai, Y., Chen, S. S., Rote, D. M., & Coffey, H. T. (1996). Vehicle/guideway dynamic interaction in maglev systems. *Journal of Dynamic Systems, Measurement and Control, 118*, 526–530.
8. Chakib, R., Essadki, A., & Cherkaoui, M. (2014). Modeling and control of a wind system based on a DFIG by active disturbance rejection control. *International Review on Modelling and Simulations (IREMOS).*
9. Cooper, K. R. (1993). Bluff-body aerodynamics as applied to vehicles. *Journal of Wind Engineering and Industrial Aerodynamics, 49*(1–3), 1–21.
10. El-Anwar, O., Ye, J., & Orabi, W. (2015). Efficient optimization of post-disaster reconstruction of transportation networks. *Journal of Computing in Civil Engineering.*
11. Elmansouri, A., El Mhamdi, J., & Boualouch, A. (2016). Wind energy conversion system using DFIG controlled by back-stepping and RST controller. In *2016 International Conference on Electrical and Information Technologies (ICEIT).*
12. Haneda, Y., Okada, M., Suganuma, Y., & Kitamura, T. (2020). A full sequence of the Matuyama–Brunhes geomagnetic reversal in the Chiba composite section, Central Japan. *Progress in Earth and Planetary Science, 7*, 44.
13. Jeong, S.-Y., Nguyen, T. H., Le, Q. A., & Lee, D.-C. (2016). High-performance control of three-phase four-wire DVR systems using feedback linearization, journal of power. *Electronics.*
14. Kamili, H., & Riffi, M. E. (2016). Portfolio optimization using the bat algorithm. *International Review on Computers and Software (IRECOS).*
15. Kerrouche, K., Mezouar, A., & Boumedien, L. (2013). A simple and efficient maximized power control of DFIG variable speed wind turbine. In *3rd international conference on systems and control.*
16. Kerrouche, K., Mezouar, A., & Boumedien, L. (2013). A simple and efficient maximized power control of DFIG variable speed wind turbine. In *3rd International Conference on Systems and Control*, Algiers, pp. 894–899. https://doi.org/10.1109/ICoSC.2013.6750963
17. Kerrouche, K. D., Mezouar, A., Boumediene, L., & K. (2014). Belgacem modeling and optimum power control based DFIG wind energy conversion system. *International Review of Electrical Engineering (I.R.E.E.), 9*(1).
18. Khodakarami, J., & Ghobadi, P. (2016). Urban pollution and solar radiation impacts. *Renewable and Sustainable Energy Reviews, 57*, 965–976.
19. Kvasnica, I., & Kvasnica, P. (2016). Multiphysics base of modelling of the MEMS micromirror and simulation by modal control. *International Review of Automatic Control (IREACO), 9*(6), 383.
20. Lala, H., & Karmakar, S. (2015). Continuous wavelet transform and artificial neural network based fault diagnosis in 52 bus hybrid distributed generation system. In *2015 IEEE Students Conference on Engineering and Systems (SCES).*
21. Lin, W., Hong, C., Huang, C. and Ou, T. (2013). Hybrid Control of a Wind Induction Generator Based on Grey–Elman Neural Network. IEEE Transactions on Control Systems Technology, 21 (6), 2367–2373.
22. Loucif, M., & Boumediene, A. (2015). Modeling and direct power control for a DFIG under wind speed variation, In *2015 3rd international conference on Control Engineering & Information Technology (CEIT).*
23. Loucif, M., & Boumédiène, A. (2015). Modeling and direct power control for a DFIG under wind speed variation. In 2015 *3rd international conference on control, Engineering & Information Technology (CEIT)*, Tlemcen, pp. 1–6. https://doi.org/10.1109/CEIT.2015.7233042
24. Lu, L. (2002). Investigation on wind power potential on Hong Kong islands-an analysis of wind power and wind turbine characteristics. *Renewable Energy.*
25. Mann, M. E., Bradley, R. S., & Hughes, M. K. (1998). Global-scale temperature patterns and climate forcing over the past six centuries. *Nature, 392*, 779–787.

26. McBeath, S. (2006). *Competition car aerodynamics: A practical handbook*. Haynes Publishing., ISBN 978-1844252305.
27. Mishra, O. P., & Naskar, D. C. (2024). *Geophysical methods—Fundamentals, applications, and case studies*. CRC Press.
28. Mone, G. (2008–2010). The driving airplane gets real. *Popular Science*, pp. 42–48. Retrieved 20 Mar 2009.
29. Pichan, M., Rastegar, H., & Monfared, M. (2013). Two fuzzy-based direct power control strategies for doubly-fed induction generators in wind energy conversion systems. *Energy, 51*, 154–162.
30. Soong, T. T. (1990). *Active structural control: Theory and practice*. Longman Scientific & Technical.
31. South, M. C. (2024). *Modified coulomb approach for squat silos*. Brigham Young University.
32. Thompson, S. J., & Congressional Research Service. (1989) *High Speed Ground Transportation (HGST): Prospects and public policy*, April 6, p. 5.
33. Yau, J. D. (2012). Lateral vibration control of a low-speed maglev vehicle in cross winds. *Wind and Structures, 15*(3), 263–283.
34. Zhao, C. F., & Zhai, W. M. (2002). Maglev vehicle/guideway vertical random response and ride quality. *Vehicle System Dynamics, 38*(3), 185–210.
35. Zheng, X. J., Wu, J. J., & Zhou, Y. H. (1997). Numerical analyses on dynamic control of five-degree-of-freedom maglev vehicle moving on flexible guideways. *Journal of Sound and Vibration, 235*, 43–61.
36. Zheng, X. J., Wu, J. J., & Zhou, Y. H. (2005). Effect of spring non-linearity on dynamic stability of a controlled maglev vehicle and its guideway system. *Journal of Sound and Vibration, 279*, 201–215.

Part VI
Conclusion

Chapter 15
The Modeling of Global Environmental Equilibrium

Introduction

The rapid growth of industrialization and the global mismanagement of natural resources have accelerated the accumulation of atmospheric CO_2 concentration and global environmental pollution [2, 7, 45]. Recently numerous studies revealed that the amount of CO_2 entering the atmosphere is currently increasing at a rate of 2.11% per year, while global environmental pollutants are increasing at a rate of 3.01% per year worldwide [12, 16, 48]. These findings suggest that the overall global environmental complexity is increasing, making it more difficult for future generations to thrive on earth [3, 11, 37]. Thus, it is imperative that we stabilize the global environment to a comfortable level by dissolving CO_2 in the atmosphere and mitigating environmental pollutants to demonstrate the planet's adaptability, versatility, manageability to ensure that the future generations can have a sustainable planet to survive [20, 21, 41]. Therefore, in this work, an effort has been made to naturally break down CO_2 to address the effects of global environmental change by implementation of phototrophic energy on the global vegetarian [18, 22, 32]. Subsequently, the activation of chemosynthesis of chemotrophs has been proposed to eradicate all inorganic and organic pollutants from the earth through the oxidation process to clean up the global environment [29, 31, 34]. This natural mechanism for the breakdown of CO_2 and eliminating global pollutants will undoubtedly be the interesting scientific discoveries of all time to naturally equilibrate the total global environmental perplexity.

© The Author(s), under exclusive license to Springer Nature Switzerland AG 2025

M. F. Hossain, *Sustainable Global Development in Environment, Energy, Building, and Infrastructure*, https://doi.org/10.1007/978-3-031-84429-4_15

Methods and Simulation

Simulation of the Phototrophic Energy ($\mathbf{Pt_e}$)

Phototrophic energy is the mechanism by which green plants absorb solar energy and transform it into chemical energy. Thus, the phototrophic energy (Pt_e) availability on the global vegetation is being computed by using the following law of cosines by taking into consideration the plane factor of the earth as follows:

$$E = \frac{hc}{\lambda} + \rho\cos(a)\cos(b) + \rho\sin(a)\sin(b)\cos(c) \tag{15.1}$$

Here, h represents the Planck constant 6.627×10^{-34} J.S, c represents the speed of light, and λ represents the photon wavelength, while the arc lengths a, b are represented the spherical triangle of the earth in radians, and C represents the angle of the vertex corresponding to the arc [8, 25, 38]. Naturally, Eq. 15.2 can be rewritten by considering the implementation of zenith angle of θ earth as follows:

$$E = \frac{hc}{\lambda} + \rho\cos(\theta) + \sin(\phi)\sin(\delta) + \cos(\phi)\cos(\delta)\cos(r) \tag{15.2}$$

To modify this calculation, a further clarification has been conducted considering a general derivative as follows:

$$E = \frac{hc}{\lambda} + \rho\cos(\theta) + \rho\sin(\phi)\sin(\delta)\cos(\beta) + \sin(\delta)\cos(\phi)\sin(\beta)\cos(\gamma) + \cos(\theta)\cos(\delta)\cos(\beta) \tag{15.3}$$

Here, β stands for the angle that corresponds to the horizontal, γ stands for the azimuth angle, and ϕ represents the spherical angle on the plane of global vegetation of earth. As a result, the phototrophic energy flux onto the global vegetation is being computed as follows:

$$Q = \begin{cases} S_o \dfrac{R_0^2}{R_E^2} \cos(\theta) & \cos(\theta) > 0 \\ 0 & \cos(\theta) \leq 0 \end{cases} \tag{15.4}$$

Hence, the global vegetation is represented by R_E and the mean distance is indicated by R_O, which stands for the average distant astronomical unit (a.u.), and S_O is the phototrophic energy constant [23, 35, 36]. Therefore, the calculation Q throughout a time is the mean value of phototrophic energy flux with respect to the time of $h = \pi$ can be rewritten as

$$Q^{\text{time}} = \frac{1}{2\pi} \int_{\pi}^{-\pi} Q dh \tag{15.5}$$

As h is the time-related angle, the equation can be solved as follows when $\phi = 1/2\pi$, or for h_O.

$$\rho \sin(\phi) \sin(\delta) + \cos(\phi) \cos(\delta) \cos(h_{0)} = 0$$

or

$$\rho \cos(h_0) = -\tan(\phi) \tan(\delta) \tag{15.6}$$

Here, $h_o = \pi$ since the phototrophic energy is being established considering its existence at $h = \pi$ thus, $\tan(\varphi)\tan(\delta) > 1$. Consequently, a Q time $= 0$ would subsequently be relatively constant over the earth's phototrophic energy emission on the global vegetation, which can be represented by considering the integral as follows:

$$\int_{\pi}^{-\pi} Q dh = \int_{h_0}^{-h_0} Q dh = S_0 \frac{R_0^2}{R_E^2} \int_{h_0}^{-h_0} \cos(\theta) dh = S_0 \frac{R_0^2}{R_E^2}$$
$$\times \left[h \sin(\phi) \sin(\delta) + \cos(\phi) \cos(\delta) \sin(h)\right]_{h = h_0}^{h = -h_0} = -2S_0 \frac{R_0^2}{R_E^2}$$
$$\times \left[h_0 \sin(\phi) \sin(\delta) + \cos(\phi) \cos(\delta) \sin\left(h_{0)}\right]\right. \tag{15.7}$$

Therefore:

$$Q^{-\text{time}} = \frac{S_0}{\pi} \frac{R_0^2}{R_E^2} \left[h_0 \sin(\phi) \sin(\delta) + \cos(\phi) \cos(\delta) \sin\left(h_{0)}\right]\right. \tag{15.8}$$

As the θ is representing here as prime angle that denotes the orbit of the earth, $\theta = 0$ considering to its vernal equinox of δ can be expressed as

$$\delta = \varepsilon \sin(\theta) \tag{15.9}$$

Here, ε denotes the prime longitude of sphere ω that is related to the vernal equinox, and thus, the equation can be addressed as

$$R_E = \frac{R_0}{1 + e \cos(\theta - \omega)} \tag{15.10}$$

or

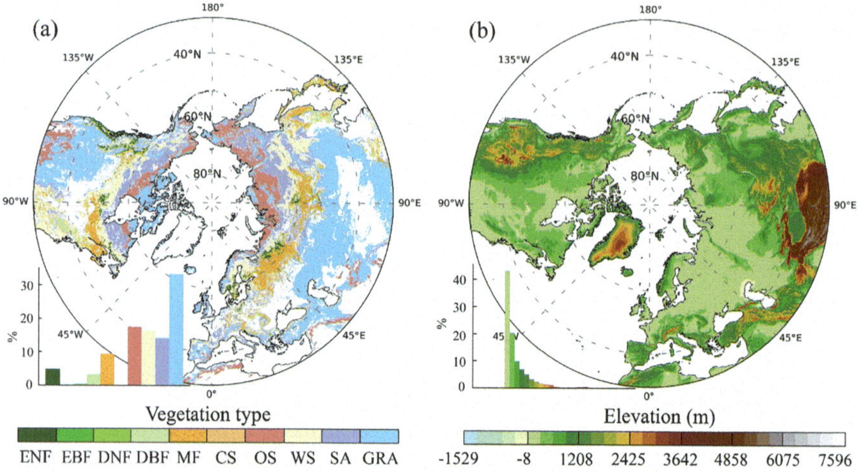

Fig. 15.1 The phototrophic energy considering (**a**) phototrophic energy flow based on vegetation type, and (**b**) phototrophic energy flow based on vegetation elevation

$$\frac{R_0}{R_E} = D_e = 1 + e\cos(\theta - \omega) \tag{15.11}$$

Consequently, the net phototrophic energy can be calculated on global vegetation for any astro-dynamical latitude on the earth at a given moment on earth as

$$D_e = S_0\left(1 + \cos\left(2\pi\frac{n}{x}\right)\right) \tag{15.12}$$

Here, S_0 represents the phototrophic energy constant respecting the sphere of the earth considering the laws of cosine, where the Planck constant, n, is the unit of time, x is the period of time that optimizes the emission rate of phototrophic energy on the global vegetation with respect vegetation type and elevation (*see* Fig. 15.1).

Implementation of Phototrophic Energy (Pt$_e$) *to Break the* CO_2

As phototrophic energy is a factual light energy, it is being addressed by initiating its quantum electrodynamics on the gauge field U(1) of the global vegetation considering its vectorial dynamics to release hypercharge electrons [1, 9, 10]. As this hypercharge electron is the striking electron force of the phototrophic energy, the release of kinetic energy proliferation on the global vegetation can be expressed as follows:

$$\overrightarrow{De} = \left[\frac{Edp}{c}, \overrightarrow{Pd}\right] + \overrightarrow{Xef}\left[\frac{Eef}{c}, \overrightarrow{Pef}\right] \tag{15.13}$$

Here, this phototrophic energy momentum on the global vegetation can be recast as follows as it is a preserved hypercharge electron:

$$\overrightarrow{De} + \overrightarrow{Xei} = \overrightarrow{Xp} + \overrightarrow{Xef} \tag{15.14}$$

Subsequently, the square of both sides of the equation can be rewritten considering the momentums of the departing hypercharge electron on the global vegetation as

$$\left(\overrightarrow{\overrightarrow{De}} + \overrightarrow{Xei} - \overrightarrow{Xp}\right)^2 = \overrightarrow{Xef}^2 \tag{15.15}$$

Consequently, the phototrophic energy hypercharge electron momentum on the global vegetation has been further analyzed computationally by multiplying its cross-term as

$$\overrightarrow{De} = 2\overrightarrow{Xd}.\overrightarrow{Xei} = 2 * \left[\frac{Ed}{c}, \overrightarrow{Pd}\right].\left[m_e c, \vec{0}\right] = 2Edm_e \tag{15.16}$$

This equation was then rewritten to represent the formation of hypercharge electrons by multiplying all the originally stationary electron's momentums on the global vegetation together with the dynamic phototrophic energy generation as follows:

$$\overrightarrow{De} = m_d^2c^2 - m_e^2c^2 + -2m_eE_p - 2\left(\frac{E_dE_p}{c^2} - \frac{P_dE_p\cos(\theta)}{c}\right) + 2E_dm_e \tag{15.17}$$

Which can be simplified as

$$\overrightarrow{De} = m_d^2c^2 - 2\left(m_eE_p + \left(\frac{E_dE_p}{c^2} - \frac{P_dE_p}{c}\cos(\theta)\right) - E_dm_e\right) \tag{15.18}$$

Subsequently, this phototrophic energy momentum on the global vegetation has been clarified by doing the following mathematical computation to validate this hypercharge electron creation as

$$\overrightarrow{De} = 2\left(m_eE_p + \left(\frac{E_dE_p}{c^2}\right) - \frac{P_dE_p}{c}\cos(\theta)\ )\right) = m_d^2c^2 - 2E_dm_e \tag{15.19}$$

Here, the generation of hypercharge electrons flow on the global vegetation is connected to the vectorial momentum of phototrophic energy; consequently, the case of scattering of phototrophic energy can be represented in the limit as $m_d \rightarrow 0$ [24, 30, 39]. Thus, the expression of the hypercharge electron implementation on the global vegetation can be measured by implementing the differential scattering of the phototrophic energy configuration, which is the striking electron force, as $m_d \rightarrow 0$, and $P_d \rightarrow E_c$ [4, 14]. As a result, the immense active force of this striking electron of the phototrophic energy would scatter the electromagnetic wave on the global vegetation [4, 26, 47]. As this hypercharge electron force is extremely sensitive to the covalent bond, it will break the covalent bond ($C = O$) of the CO_2 to transform it into C and O_2 [11, 24, 27]. As a result, the equation can be expressed as the derivative process of the hypercharge striking electron force of the phototropic energy on the global vegetation that is highly kinematic with respect to electromagnetic waves:

$$\overrightarrow{De} = 2\left[\frac{Edp}{c}, \overrightarrow{Pd}\right] + \overrightarrow{CO2} + \left[\overrightarrow{Pemf}\right] \tag{15.20}$$

$$= 2\left[\frac{Edp}{c}, \overrightarrow{Pd}\right] + \left[\overrightarrow{C = O2}\right] - [\text{Covalent Bond}] \tag{15.21}$$

As phototrophic energy momentum on the global vegetation is hypercharge, momentums of the electron can be considered the chemical energy as

$$\overrightarrow{De} = 2\left[\frac{Ed}{c}, \overrightarrow{Pd}\right] + \left[\overrightarrow{C + O2}\right] - \left[m_e\overrightarrow{c^2}\right] = 2Edm_e \tag{15.22}$$

Here, $\overrightarrow{De} = S_0\left(1 + \cos\left(2\pi\frac{n}{x}\right)\right)$ and, thus, this CO_2 can be broken down by implementing first-order hypercharged electron of the phototropic energy momentums that can be expressed as.

$$\overrightarrow{De} = 2E_d m_e = S_0\left(1 + \cos\left(2\pi\frac{n}{x}\right)\right) - [C + O_2] \tag{15.23}$$

Simulation of the Chemotrophic Mechanism (Ct_m)

The chemotroph obtains food by the oxidation of electron donors in their environmental pollutants of both inorganic and organic compounds; therefore, the global pollutants-eliminating mechanisms have been clarified concerning the common *metric signature* convention of chemotrophs (*see* Fig. 15.2).

Hence, the net computation of the global total pollutant has been calculated by cumulating both inorganic and organic pollutant compounds presence on earth by expressing Eq. 15.24:

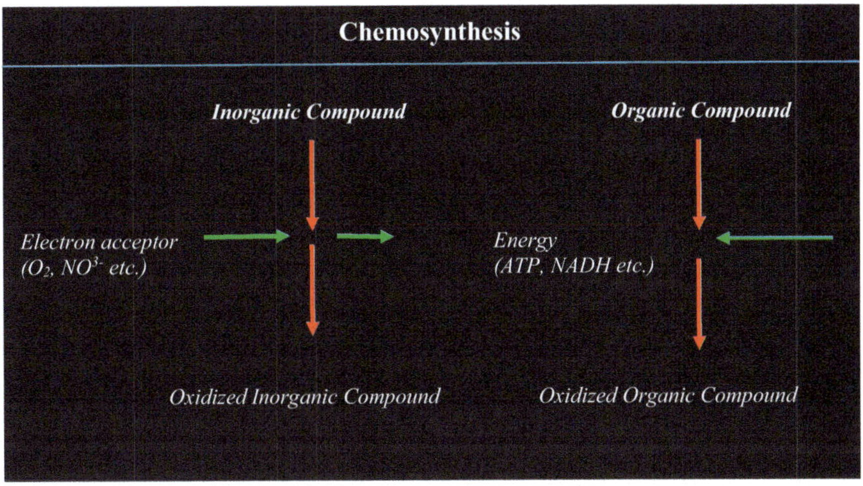

Fig. 15.2 The schematic mechanism of the chemosynthesis by the chemotrophs to take their from both inorganic and organic pollutants from the environment though the oxidation process

$$\left[\nabla^2 = \frac{i^2 o^2}{t^2}\right]\psi(r) \tag{15.24}$$

Here, ∇^2 represents the earth plane, while i represents the pollutant from the inorganic compound, o represents the pollutants from the organic compounds, and t represents the time function at ψ with respect to the areal radius of the earth r.

As the earth plane has a standard unit, $(c + m^2)\psi(x) = 0$, thus, the metric signature of $h^{io} = (\pm 1D, \mp 2D, \mp 3D)$ can be clarified as the net global pollutants as

$$\psi(x) = \int \frac{d^4 p}{(2\pi)^4} e^{-ip.x}\psi(p) \tag{15.25}$$

and using 3D dimension of the earth, the net inorganic and organic pollutants can be further described as

$$\psi(x) = \int \frac{d^4 p}{(2\pi)^4} e^{ip.x} C(P)\delta\left(P^0\right)^2 - E(P)^2\big) \tag{15.26}$$

As it is the standard function to accurately compute the total pollutants presence on earth, mathematical process of chemotrophic mechanism can be described as

$$\psi(x) = \int \frac{d^4 p}{(2\pi)^4} \delta\left(\left(p^0\right)^2 - E(p)^2\right)\left(A(p)e^{-ip^0x^0 + ip^ix^i} + B(p)e^{+ip^0x^0 + ip^ix^i}\right)\theta\left(p^0\right) \tag{15.27}$$

$$= \int \frac{d^4p}{(2\pi)^4} \delta\left((p^0)^2 - E(p)^2 \right) \left(A(P) e^{-ip^0x^0 + ip^ix^i} + B(-P) e^{+ip^0x^0 - ip^ix^i} \right) \theta(p^0)$$
$$\rightarrow \int \frac{d^4p}{(2\pi)^4} \delta\left((p^0)^2 - E(p)^2 \right) \left(A(p) e^{-ip.x} + B(P) e^{+ip.x} \right) \theta(p^0) \tag{15.28}$$

Chemotrophic Mechanism to Eliminate Pollutants

The global net chemotrophs are being further analyzed to eliminate all pollutants from the earth by process of oxidation through chemosynthesis of the chemotrophs [5, 19, 43]. Thus, the global plane of local symmetry of earth is being clarified to initiate the oxidation process of all pollutants present on the earth. Subsequently, the earth plane has been modeled with Abelian local symmetries scalar field computation of all chemotrophs of the earth to confirm that they are energetically oxidized by engulfing all pollutants by expressing Eq. 15.29:

$$Ct = F_{\mu\upsilon} F^{\mu\upsilon} + D_\mu \Phi^* D^\mu \Phi - V(\Phi^* \Phi) \tag{15.29}$$

where

$$\begin{aligned} D_\mu \Phi(x) &= \partial_\mu \Phi(x) + \mathrm{iqA_\mu(x)\Phi\ (x)} \\ D_\mu \Phi^*(x) &= \partial_\mu \Phi^*(x) - \mathrm{iqA_\mu(x)\Phi^*(x)}, \end{aligned} \tag{15.30}$$

and

$$V(\Phi^* \Phi) = \frac{\lambda}{2} (\Phi^* \Phi)^2 + m^2 (\Phi^* \Phi). \tag{15.31}$$

Here, $\lambda > 0$ and $m^2 < 0$ So, $\Phi = 0$ is the total scalar field of the global environment. As a result, the scalar field of the chemotrophs (x) will address the earth's local U(1) symmetries to eat all pollutants which can be expressed by

$$\Phi\ (\mathrm{x}) = \frac{1}{\sqrt{2}} \Phi_\mathrm{r}(\mathrm{x}) * \mathrm{e}^{\mathrm{i}\Theta(\mathrm{x})}, \text{real } \Phi_\mathrm{r}(\mathrm{x}) > 0, \text{real } \Phi(\mathrm{x}). \tag{15.32}$$

Here, the scalar field of chemotrophs satisfies the oxidation process to eat all pollutants on earth that can be expressed as

$$V(\phi) = \frac{\lambda}{8} \left(\phi_r^2 - v^2 \right)^2 + \text{const.} \tag{15.33}$$

Subsequently, the net environmental cleanup due to the momentum field of the oxidation process by chemotrophs can be shifted by applying the following variable scalars, $\Phi_r(x) = v + \sigma(x)$, and expressed as

$$\phi_r^2 - v^2 = (v + \sigma)^2 - v^2 = 2v\sigma + \sigma^2 \tag{15.34}$$

$$V = \frac{\lambda}{8} \left(2v\sigma - \sigma^2\right)^2 = \frac{\lambda v^2}{2} * \sigma^2 + \frac{\lambda v}{2} * \sigma^3 + \frac{\lambda}{8} * \sigma^4. \tag{15.35}$$

Thus, the functional derivative $D_\mu \phi$ will become

$$D_\mu \phi = \frac{1}{\sqrt{2}} \left(\partial_\mu \left(\phi_r e^{i\Theta}\right) + iqA_\mu * \phi_r e^{i\Theta} \right) = \frac{e^{i\Theta}}{\sqrt{2}} \left(\partial_\mu \phi_r + \phi_r * i\partial_\mu \Theta + \phi_r * iqA_\mu\right) \tag{15.36}$$

$$= \frac{1}{2} \left(\partial_\mu \sigma\right)^2 + \frac{(v + \sigma)^2}{2} * \left(\partial_\mu \Theta + qA_\mu\right)^2 \tag{15.37}$$

Simply, this oxidation process to fulfill the nutritional requirements of the chemotrophs will eat all pollutants from the earth; thus, it will clean up the global environmental pollutants completely.

Results and Discussion

Simulation of the Phototrophic Energy (Pt$_e$)

The resultant clarification of the phototrophic energy is being computed considering gauge field U(1) of the global vegetation considering the earth plane as

$$D_e = P_{pv} = \eta_{pvg} \, A_{pvg} \, G_t. \tag{15.38}$$

Here, the phototrophic energy generation rate is represented by ηpvg, the gauge field is represented by Apvg (m^2), and the phototrophic energy intake rate on the global vegetation is represented by Gt (W/m^2). As a result, the ηpvg can be recast as follows:

$$\eta_{pvg} = \eta_r \eta_{pc} \left[1 - \beta\left(T_c - T_{c\ ref}\right)\right] \tag{15.39}$$

Here, β stands for the time cofactor, ηr stands for the phototrophic energy efficiency mode, and T*cref* stands for the tangential condition in consideration of the global vegetation, and thus, the equation can be rewritten as follows:

$$T_c = T_a + \left(\frac{NOCT - 20}{800}\right) G_t \tag{15.40}$$

Given that T_a here represents the encircling tangential and G_t represents the phototrophic energy generation on the global vegetarian (W/m^2). Thus, Eq. 15.41 is to be used to determine the accurate phototrophic energy release on the global vegetation:

$$I_t = I_b R_b + I_d R_d + (I_b + I_d) R_r \tag{15.41}$$

Subsequently, the phototrophic energy throughout the global vegetation, which is the electron charge (I) produced from the gauge field U(1) on the earth plane, also calculated by using Eq. 15.42:

$$I = N_p \left[I_{ph} - I_{rs} \left[\exp\left(\frac{q(V + IR_s)}{AKTN_s} - 1 \right) \right] \right] \tag{15.42}$$

where

$$I_{rs} = I_{rr} \left(\frac{T}{T_r} \right)^3 \exp\left[\frac{E_G}{AK} \left(\frac{1}{T_r} - \frac{1}{T} \right) \right] \tag{15.43}$$

Here, the electron charge ($q = 1.6 \times 10^{-19}$ C) is being computed by the standardized efficiency (AK) of the global plane of the considering the tangential T [40, 42, 46]. Hence, T_r indicates the earth plane condition in relation to the global vegetation, I_{rr} indicates the reversed photrophic energy at T_r, and E_G indicates the phototrophic energy rate released on the global vegetation; the total phototrophic energy I_{ph} release on the global vegetation can be expressed as follows:

$$I_{ph} = \left[I_{SCR} + k_i (T - T_r) \frac{S}{100} \right] \tag{15.44}$$

Hence, the accurate computational model of the phototrophic energy generation is being represented by considering the optimum energy radiation dynamic on global vegetation at rate of unit area (W/m^2).

$$I = I_{ph} - I_D \tag{15.45}$$

$$I = I_{ph} - I_0 \left[\exp\left(\frac{q(V + R_s I)}{AKT} - 1 \right) \right] \tag{15.46}$$

Consequently, Eq. 15.47 can be rewritten to calculate the phototrophic energy flow into the global vegetation as

$$I = I_{PV} - I_{D1} - \left(\frac{V + IR_S}{R_{SH}}\right) \tag{15.47}$$

where

$$I_{D1} = I_{01}\left[\exp\left(\frac{V + IR_s}{a_1 V_{T1}}\right) - 1\right] \tag{15.48}$$

In this case, V_{T1} and V_{T2} stand for the phototrophic energy situation, while I and I_{01} represent the reversed phototrophic energy flow into the global vegetation vector [6, 8, 13]. As a result, the vector of global vegetation is represented by a_1 to validate the net phototrophic energy generation on any array of global vegetation by expressing Eq. 15.49:

$$v_{oc} = \frac{V_{oc}}{cK\,T/q} \tag{15.49}$$

$$P_{\max} = \frac{\frac{V_{oc}}{cK\,T/q} - \ln\left(\frac{V_{oc}}{cK\,T/q} + 0.72\right)}{\left(1 + \frac{V_{oc}}{K\,T/q}\right)}\left(1 - \frac{V_{oc}}{\frac{V_{oc}}{I_{SC}}}\right)\left(\frac{V_{oc0}}{1 + \beta \ln \frac{G_0}{G}}\right)\left(\frac{T_o}{T}\right)^{y} I_{sc0}\left(\frac{G}{G_o}\right)^{a} \tag{15.50}$$

where T denotes the tangential of the earth's plane, K stands for Boltzmann's constant, v_{oc} stands for the standard point of the phototrophic energy flow, v_{oco} stands for the phototrophic energy condition $Vt = nkT/q$, c stands for phototrophic energy motion, α stands for the function representing the nonlinear motion of phototrophic energy, q stands for electron charge, γ stands for the function acting for all nonlinear phototrophic energy, and β stands for the global vegetation.

Implementation of Phototrophic Energy (Pt$_e$) *to Break the* CO_2

The functional test of the global Albanian symmetries of the global vegetation has been analyzed to implement phototrophic energy to break the CO_2 naturally [28, 33, 34]. Subsequently, the computed results of the phototrophic energy on the global symmetrical array indicated that phototrophic energy delivered on the global vegetation is a complex vector field that in fact a phototrophic energy charge q related to it electromagnetic field which can be expressed as

$$P_{\max} = \frac{\frac{V_{oc}}{cK\,T/q} - \ln\left(\frac{V_{oc}}{cK\,T/q} + 0.72\right)}{\left(1 + \frac{V_{oc}}{K\,T/q}\right)} \left(1 - \frac{V_{oc}}{\frac{V_{oc}}{I_{SC}}}\right) \left(\frac{V_{oc0}}{1 + \beta \ln \frac{G_0}{G}}\right) \left(\frac{T_o}{T}\right)^y I_{sc0} \left(\frac{G}{G_o}\right)^a \tag{15.51}$$

$$\mathfrak{H} = -\frac{1}{4} F_{\mu\nu} F^{\mu\nu} + D_\mu \Phi^* D^\mu \Phi - V(\Phi^* \Phi) \tag{15.52}$$

Where

$$D_\mu \Phi(x) = \partial_\mu \Phi(\mathrm{x}) + \mathrm{iqA_\mu(x)}\Phi\,(\mathrm{x}) \tag{15.53}$$

$$D_\mu \Phi^*(x) = \partial_\mu \Phi^*(\mathrm{x}) - \mathrm{iqA_\mu(x)}\Phi^*(\mathrm{x}) \tag{15.54}$$

And

$$V(\Phi^* \Phi) = \frac{\lambda}{2}\,(\Phi^* \Phi)^2 + m^2(\Phi^* \Phi) \tag{15.55}$$

Here, $\lambda > 0\; m^2 < 0$, thus,$\Phi = 0$ is a local maximum vector with minimum form of degenerated scalar of $\Phi = \frac{v}{\sqrt{2}} * e^{i\theta}$,

$$v = \sqrt{\frac{-2m^2}{\lambda}}, \text{ any real } \theta \tag{15.56}$$

Subsequently, the global vegetation will be impacted by the global U (1) net symmetrical phototrophic energy that can be stated by Eq. 15.57:

$$\Phi\,(\mathrm{x}) = \frac{1}{\sqrt{2}}\,\Phi_\mathrm{r}(\mathrm{x}) * \mathrm{e}^{\mathrm{i}\Theta(\mathrm{x})}, \text{real } \Phi_\mathrm{r}(\mathrm{x}) > 0, \text{real } \Phi(\mathrm{x}) \tag{15.57}$$

As the vector (x) = 0 is a first-order function, the hypercharge energy generation from phototrophic energy on the global vegetation will initiate to break the atmospheric CO_2 [21, 27, 30]. Naturally, Eq. 15.58 has been used to confirm the net hypercharge energy released from the phototrophic energy on the global vegetation to break the CO_2 naturally from the air:

$$V(\phi) = \frac{\lambda}{8}\,\left(\phi_r^2 - v^2\right)^2 + \text{const}, \tag{15.58}$$

or the resultant hypercharge energy generation is shifted as $\Phi_\mathrm{r}(\mathrm{x}) = \mathrm{v} + \sigma(\mathrm{x})$,

$$\phi_r^2 - v^2 = (v+\sigma)^2 - v^2 = 2v\sigma + \sigma^2 \tag{15.59}$$

$$V = \frac{\lambda}{8}\left(2v\sigma - \sigma^2\right)^2 = \frac{\lambda v^2}{2} * \sigma^2 + \frac{\lambda v}{2} * \sigma^3 + \frac{\lambda}{8} * \sigma^4 \tag{15.60}$$

Subsequently, the functional derivative $D_\mu \phi$ will become

$$D_\mu \phi = \frac{1}{\sqrt{2}}\left(\partial_\mu\left(\phi_r e^{i\Theta}\right) + iqA_\mu * \phi_r e^{i\Theta}\right) = \frac{e^{i\Theta}}{\sqrt{2}}\left(\partial_\mu \phi_r + \phi_r * i\partial_\mu \Theta + \phi_r * iqA_\mu\right) \tag{15.61}$$

$$\begin{aligned}\left|D_\mu \phi\right|^2 &= \frac{1}{2}\left|\partial_\mu \phi_r + \phi_r * i\partial_\mu \Theta + \phi_r * iqA_\mu\right|^2 = \frac{1}{2}\left(\partial_\mu \phi_r\right) \\ &+ \frac{\phi_r^2}{2} * \left(\partial_\mu \Theta qA_\mu\right)^2 = \frac{1}{2}\left(\partial_\mu\, \sigma\right)^2 + \frac{(v+\sigma)^2}{2} * \left(\partial_\mu \Theta + qA_\mu\right)^2\end{aligned} \tag{15.62}$$

Altogether,

$$\mathfrak{h} = \frac{1}{2}\left(\partial_\mu\, \sigma\right)^2 - v\,(\sigma) - \frac{1}{4}\, F_{\mu\nu}F^{\mu\nu} + \frac{(v+\sigma)^2}{2} * \left(\partial_\mu \Theta + qA_\mu\right)^2 \tag{15.63}$$

Hence, the hypercharge ($\mathfrak{h}$) energy is very sensitive to CO_2 as it is an integrated realistic vector of the phototrophic energy, and thus, the equation can be written as

$$\begin{aligned}\mathfrak{h}_{sef} &= \frac{1}{2}\left(\partial_\mu\, \sigma\right)^2 - \frac{\lambda v^2}{2} * \sigma^2 - \frac{1}{4}\, F_{\mu\nu}F^{\mu\nu} + \frac{v^2}{2} * \left(qA_\mu + \partial_\mu \Theta\right)^2 \\ &= 2\left[\frac{Ed}{c}, \overrightarrow{Pd}\right] + \left[\overrightarrow{CO_2}\right] - \left[\overrightarrow{fc^2}\right]\end{aligned} \tag{15.64}$$

So,

$$\mathfrak{h}_{sef} = 2\left[\frac{Ed}{c}, \overrightarrow{Pd}\right] + \left[\overrightarrow{CO_2}\right] - \left[\overrightarrow{fc^2}\right] \tag{15.65}$$

As $\mathfrak{h}_{sef} = De$, the equation finally can be written as

$$D_e = 2\left[\frac{Ed}{c}, \overrightarrow{Pd}\right] + \left[\overrightarrow{C + O_2}\right] - \left[\overrightarrow{f_e c^2}\right] \tag{15.66}$$

This finding indicated that phototrophic energy is indeed the crucial element in breaking the atmospheric CO_2 due to the hypercharge energy momentum on the global vegetation fields (x) that will confirm to significantly mitigate the global CO_2 emission.

Simulation of the Chemotrophic Mechanism (Ct$_m$)

By calculating its energy by the oxidation of electron donors from their environments of the chemotrophs, the per unit areal rate of chemotrophic mechanism has been clarified computationally on a global scale by using Eq. 15.67:

$$(c+\mu^2)\varphi(x)=0 \tag{15.67}$$

Then, multiplying by the chemotroph variables of $\varphi^*(x)$ and $\varphi(x)$, it can be expressed as

$$\varphi^*(c+\mu^2)\varphi=0 \tag{15.68}$$

$$\varphi(c+\mu^2)\varphi^*=0 \tag{15.69}$$

Here, mechanism of the chemotrophs on the global environmental pollutants can finally be addressed as

$$T^{\mu\nu}=\frac{\hbar^2}{m}\left(\eta^{\mu\alpha}\eta^{\nu\beta}+\eta^{\mu\beta}\eta^{\nu\alpha}-\eta^{\mu\nu}\eta^{\alpha\beta}\right)\partial_\alpha\overline{\psi}\partial_\beta\psi-\eta^{\mu\nu}mc^2\overline{\psi}\psi \tag{15.70}$$

As the clarity of global environment pollutants is related to the qualitative form of chemotrophic mechanism, the oxidation of electron donors of all chemotrophs has been calculated on global scale as follows

$$J(\omega)=\varrho(\omega)|V(\omega)|^2. \tag{15.71}$$

Hence, ω_c represents the oxidation frequency of the chemotrophs; here, the donation of the electrons $\langle a(t)\rangle = u(t,t_0)\langle a(t_0)\rangle$ is being described by clarifying the integral–differential equation as

$$u(t,t_0)=\frac{1}{1-\Sigma'(\omega_b)}e^{-i\omega(t-t_0)}+\int_{\omega_e}^{\infty}d\omega\frac{J(\omega)e^{-i\omega(t-t_0)}}{[\omega-\omega_c-\Delta(\omega)]^2+\pi^2J^2(\omega)}, \tag{15.72}$$

where $\Sigma'(\omega_b)=[\partial\Sigma(\omega)/\partial\omega]_{\omega=\omega_b}$ and $\Sigma(\omega)$ are the oxidation induction rate as

$$\Sigma(\omega)=\int_{\omega_e}^{\infty}d\omega'\frac{J(\omega')}{\omega-\omega'}. \tag{15.73}$$

Thus, the election donation mode is being computed as the primary-valued integral of the oxidation dynamics (eV) that will eventually reach a nonequilibrium oxidation state to initiate eliminate the global pollutants naturally that can be expressed as follows:

$$\rho(t) = \sum_{n=0}^{\infty} \mathcal{P}_n^{(n_0)}(t)|n_0\rangle\langle n_0| \tag{15.74}$$

Chemotrophic Mechanism to Eliminate Pollutants

Finally, this mechanism is being confirmed computationally by the following variable analysis of the chemotrophs by taking into consideration the specific scalar field of the global environmental pollutants [14, 15, 44]. As the global environmental pollutants are consisting of the local symmetry of U(1) global earth plane, it will allow to induct oxidation rapidly in terms of taking electrons from the pollutants to clean up the environment that can be expressed as [1, 17, 41].

$$\begin{aligned} &\partial_\mu \rightarrow D_\mu = \partial_\mu = ieA_\mu \quad [\text{covariant oxidations}] \\ &A'_\mu = A_\mu + \frac{1}{e}\,\partial_\mu\alpha \quad [A_\mu \text{ global oxidations}] \end{aligned} \tag{15.75}$$

Here, the local U(1) gauge invariant is being considered the perplex scalar field of the global pollutants that is expressed by

$$\mathcal{L} = (D^\mu)^\dagger\,(D_\mu\varnothing) - \frac{1}{4}\,F_{\mu\nu}F^{\mu\nu} - V(\varnothing). \tag{15.76}$$

This kinetic term of oxidation is the kinetic motion of the electrons of the global pollutants; therefore, the chemotrophic mechanism to eliminate the global pollutants can be written as

$$\mathcal{L}_{kin}(\eta,\xi) = (D^\mu\phi)^\dagger(D^\mu\phi) = (\partial^\mu + ieA^\mu)\phi^*\,(\partial_\mu - ieA_\mu)\,\phi \tag{15.77}$$

Here, $V(\eta, \xi) = \lambda\, \upsilon^2\eta^2$ is the final term of the oxidation mechanism to the second order, and thus, the chemotrophic mechanism can be expressed as

$$\mathcal{L}_{kin}(\eta,\xi) = \frac{1}{2}\,(\partial_\mu\eta)^2 - \lambda\upsilon^2\eta^2 + \frac{1}{2}\,(\partial_\mu\xi)^2 - \frac{1}{4}\,F_{\mu\nu}F^{\mu\nu} + \frac{1}{2}\,e^2\upsilon^2A_\mu^2 - e\upsilon A_\mu(\partial^\mu\xi) \tag{15.78}$$

Here, ξ denotes the electrons and μ denotes the rate of electrons. As a result, the precise term of chemotrophic mechanism to address the global pollutant can be expressed as:

$$\mathcal{L}_{\text{scalar}} = (D^{\mu}\phi)^{\dagger}(D^{\mu}\phi) - V(\phi^{\dagger}\phi)$$
$$= (\partial^{\mu} + ieA^{\mu})\frac{1}{\sqrt{2}}(v+h)(\partial_{\mu} - ieA_{\mu})\frac{1}{\sqrt{2}}(v+h) - V(\phi^{\dagger}\phi) \quad (15.79)$$

$$= \frac{1}{2}(\partial_{\mu}h)^{2} + \frac{1}{2}e^{2}A_{\mu}^{2}(v+h)^{2} - \lambda v^{2}h^{2} - \lambda vh^{3} - \frac{1}{4}\lambda h^{4} + \frac{1}{4}\lambda h^{4} \quad (15.80)$$

Hence, this calculative chemotrophic mechanism validates the net oxidation representation, which ascertains the net electron donation of s0 in terms of elimination of global pollution from their environment that can be expressed as.

$$\overline{\phi}[s_0] = 2s_0(\ln 4s_0 - 2) + \ln 4s_0(\ln 4s_0 - 2) - \frac{(\pi^2 - 9)}{3} + s_0^{-1}\left(\ln 4s_0 + \frac{9}{8}\right)$$
$$+ \cdots (s_0 \gg 1); \quad (15.81)$$

$$\overline{\phi}[s_0] = \left(\frac{2}{3}\right)(S_0 - 1)^{\frac{3}{2}} + \left(\frac{5}{3}\right)(S_0 - 1)^{\frac{5}{2}} - \left(\frac{1507}{420}\right)(S_0 - 1)^{\frac{7}{2}}$$
$$+ \cdots (s_0 - 1 \ll 1) \quad (15.82)$$

As the function is $1 < s0 < 10$, the chemotrophic mechanism can be further calculated precisely using oxidation rate at the global environmental pollutants as follows:

$$n(\in) = 0, \in < \in_0 = C_{\in}^{-\alpha} \text{ or } D_{\in}^{\beta}, \in_0 < \in < \in_m = 0, \in > \in_m \quad (15.83)$$

Then, it can be transformed as

$$\left(\frac{d\tau_{abs}}{dx}\right)_{\alpha} = \pi r_0^2 C\left(\frac{m^2c^4}{E}\right)^{1-\alpha} \times \begin{cases} 0, E < E_m \\ [F_{\alpha}(1) - F_{\alpha}(\sigma_m)], E_m < E < E_0 \\ [F_{\alpha}(\sigma_0) - F_{\alpha}(\sigma_m)], E > E_0; \end{cases} \quad (15.84)$$

$$\left(\frac{d\tau_{abs}}{dx}\right)_{\beta} = \pi r_0^2 D\left(\frac{m^2c^4}{E}\right)^{1+\beta} \times \begin{cases} 0, E < E_m \\ [F_{\beta}(\sigma_m), E_m < E < E_0 \\ [F_{\beta}(\sigma_m) - F_{\beta}(\sigma_0)], E > E_0. \end{cases} \quad (15.85)$$

Consequently, this equation is being simplified by applying the contributed chemotrophic mechanism rate to computationally mitigate the global net pollutants as

$$\frac{d\Gamma_{\gamma}^{LPM}}{d^3k} = \frac{d_F q_s^2 \alpha_{EM}}{4\pi^2 k} \int_{-\infty}^{\infty} \frac{dp_{\|}}{2\pi} \int \frac{d^2 \mathrm{p}_{\perp}}{(2\pi)^2} A\left(p_{\|}, k\right) \, \mathrm{Re}\left\{2\mathrm{P}_{\perp} \cdot f\left(\mathrm{p}_{\perp}; p_{\|}, k\right)\right\} \quad (15.86)$$

where d_{F} is the variable state of the global pollutants [N_c in SU(N_c)], q_s is the charge of the oxidations, k ≡ |k|. Thus, the net kinetic function of chemotrophs has been computed to eliminate global pollutant A(p||, k) as

$$A\left(p_{\|}, k\right) \equiv \begin{cases} \dfrac{n_b\left(k + p_{\|}\right)\left[1 + n_b\left(p_{\|}\right)\right]}{2p_{\|}\left(p_{\|} + k\right)} \\ \dfrac{n_f\left(k + p_{\|}\right)\left[1 - n_f\left(p_{\|}\right)\right]}{2\left[p_{\|}\left(p_{\|} + k\right)\right]^2} \left[p_{\|}^2 + \left(p_{\|} + k\right)^2\right] \end{cases} \quad (15.87)$$

Conclusions

By using phototrophic energy and chemotrophic mechanisms to balance global CO_2 and the environmental pollutants, a comprehensive analysis has been performed to secure a sustainable planet. Hence, the application of quantum electrodynamics of the phototrophic energy of the lights has been used to release hypercharge electrons on the global vegetation to naturally break down CO_2. This is because the hypercharge electron is the striking electron force of the phototrophic energy that releases kinetic energy to break the atmospheric CO_2 rapidly. Subsequently, all pollutants from the earth through the chemosynthesis of the chemotrophs have been modeled to eliminate all inorganic and organic pollutants from the earth through process of its oxidation mechanism. Simply, using phototrophic energy to break the atmospheric CO_2 and implementing the chemosynthesis process of chemotrophs to naturally clean up the global environmental pollutants shall indeed be a great discovery in natural science to ensure a sustainable plane for the future generation.

Acknowledgments This research has no conflict of interest to publish any suitable journal or publisher.

References

1. Artemyev, N., Jentschura, U., Serbo, V., & Surzhykov, A. (2012). Strong electromagnetic field effects in ultra-relativistic heavy-ion collisions. *European Physical Journal C: Particles and Fields, 72*, 1935.

2. Baur, G., Hencken, K., Trautmann, D., Sadovsky, S., & Kharlov, Y. (2002). Dense laser-driven electron sheets as relativistic mirrors for coherent production of brilliant X-ray and γ-ray beams. *Physics Reports, 364*, 359–450.
3. Baur, G., Hencken, K., & Trautmann, D. (2007). Revisiting unitarity corrections for electromagnetic processes in collisions of relativistic nuclei. *Physics Reports, 453*, 1–27.
4. Birnbaum, K. M., et al. (2005). Photon blockade in an optical cavity with one trapped atom. *Nature, 436*, 87–90.
5. Boettcher, I., Pawlowski, J. M., & Diehl, S. (2012). Ultracold atoms and the functional renormalization group. *Nuclear Physics B - Proceedings Supplements, 228*, 63–135.
6. Broz, M., et al. (2020). A generator of forward neutrons for ultra-peripheral collisions. *Computer Physics Communications*.
7. Busch, K., et al. (2007). Periodic nanostructures for photonics. *Physics Reports, 444*, 101–202.
8. Cardoso, V., Lemos, J. P., & Yoshida, S. (2004). Quasinormal modes of Schwarzschild black holes in four and higher dimensions. *Physical Review D, 69*, 044004.
9. Cartwright, J. (2010). Photons meet with three-way split. *Nature*. https://doi.org/10.1038/news.2010.381
10. Chang, D. E., Sørensen, A. S., Demler, E. A., & Lukin, M. D. (2007). A single-photon transistor using nanoscale surface plasmons. *Nature Physics, 3*, 807–812.
11. Chen, J., Wang, C., Zhang, R., & Xiao, J. (2012). Multiple plasmon-induced transparencies in coupled-resonator systems. *Optics Letters, 37*, 5133–5135.
12. Chen, G., et al. (2013). Examining non-locality and quantum coherent dynamics induced by a common reservoir. *Scientific Reports*.
13. Cheng, M. T., & Song, Y. Y. (2012). Fano resonance analysis in a pair of semiconductor quantum dots coupling to a metal nanowire. *Optics Letters, 37*, 978–980.
14. Dayan, B., et al. (2008). A photon turnstile dynamically regulated by one atom. *Science, 319*, 1062–1065.
15. Destrade, M. (2003). Elastic interface acoustic waves in twinned crystals. *International Journal of Solids and Structures*.
16. Dobrynina, A., Kartavtsev, A., & Raffelt, G. (2015). Photon-photon dispersion of TeV gamma rays and its role for photon-ALP conversion. *Physical Review D, 91*, 083003.
17. Douglas, J. S., et al. (2015). Quantum many-body models with cold atoms coupled to photonic crystals. *Nature Photonics, 9*, 326–331.
18. Dupuis, N., Canet, L., Eichhorn, A., Metzner, W., Pawlowski, J. M., Tissier, M., & Wschebor, N. (2021). The nonperturbative functional renormalization group and its applications. *Physics Reports*.
19. Dupuis, N. L., et al. (2021). The nonperturbative functional renormalization group and its applications. *Physics Reports*.
20. Eichler, J., & Stöhlker, T. (2007). Radiative electron capture in relativistic ion–atom collisions and the photoelectric effect in hydrogen-like high-Z systems. *Physics Reports, 439*, 1–99.
21. Englund, D., et al. (2010). Resonant excitation of a quantum dot strongly coupled to a photonic crystal nanocavity. *Physical Review Letters, 104*, 073904.
22. Fernandez, J., & Martín, F. (2009). Electron and ion angular distributions in resonant dissociative photoionization of H2 and D2 using linearly polarized light. *New Journal of Physics, 11*, 043020.
23. Gabor, M., et al. (2009). Extremely efficient multiple electron-hole pair generation in carbon nanotube photodiodes. *Science, 325*, 1367–1371.
24. Gleyzes, S., et al. (2007). Quantum jumps of light recording the birth and death of a photon in a cavity. *Nature, 446*, 297–300.
25. Gould, R. J. (1967). Pair production in photon-photon collisions. *Physics Review, 155*, 1404–1406.
26. Guerlin, C., et al. (2007). Progressive field-state collapse and quantum non-demolition photon counting. *Nature, 448*, 889–893.

27. Guo, Y., Al-Jubainawi, A., & Ma, Z. (2019). Performance investigation and optimisation of electrodialysis regeneration for LiCl liquid desiccant cooling systems. *Applied Thermal Engineering, 149*, 1023–1034.
28. Hencken, K., Baur, G., & Trautmann, D. (2006). Transverse momentum distribution of vector mesons produced in ultraperipheral relativistic heavy ion collisions. *Physical Review Letters, 96*, 012303.
29. Hübel, H., et al. (2010). Direct generation of photon triplets using cascaded photon-pair sources. *Nature, 466*, 601–603.
30. Igor, B., et al. (2012). Ultracold atoms and the functional renormalization group. *Nuclear Physics B - Proceedings Supplements.*
31. Javadi, A., et al. (2015). Single-photon non-linear optics with a quantum dot in a waveguide. *Nature Communications, 6*, 8655.
32. Jiaqi, L., Yi, Z., Chengkai, T., & Zhu, X. (2019). *INS aided high dynamic single-satellite position algorithm.* In 2019 IEEE international conference on signal processing, communications and computing (ICSPCC).
33. Johnson, B. R. (2010). Quantum non-demolition detection of single microwave photons in a circuit. *Nature Physics.*
34. Kaneda, F., et al. (2019). High-efficiency single-photon generation via large-scale active time multiplexing. *Science Advances, 5*(10), eaaw8586.
35. Kevin, A., et al. (2017). Signatures of two-photon pulses from a quantum two-level system. *Nature Physics, 3*, 649–654.
36. Langer, L., et al. (2014). Access to long-term optical memories using photon echoes retrieved from semiconductor spins. *Nature Photonics, 8*, 851–857.
37. Langford, K., et al. (2021). Efficient quantum computing using coherent photon conversion. *Nature, 478*, 360–363.
38. Naghiloo, M., et al. (2016). Mapping quantum state dynamics in spontaneous emission. *Nature Communications.*
39. Najjari, B., Voitkiv, A. B., Artemyev, A., & Surzhykov, A. (2009). Simultaneous electron capture and bound-free pair production in relativistic collisions of heavy nuclei with atoms. *Physical Review A, 80*, 012701.
40. Prasad, A. S., Hinney, J., Mahmoodian, S., Hammerer, K., et al. (2020). Correlating photons using the collective nonlinear response of atoms weakly coupled to an optical mode. *Nature Photonics.*
41. Reinhard, A., et al. (2012). Strongly correlated photons on a chip. *Nature Photonics, 6*, 93–96.
42. Reinhard, P., et al. (2021). Nuclear charge densities in spherical and deformed nuclei: Toward precise calculations of charge radii. *Physical Review C.*
43. Shalm, L. K., et al. (2012). Three-photon energy–time entanglement. *Nature Physics.*
44. Stehle, C., Zimmermann, C., & Slama, S. (2014). Cooperative coupling of ultracold atoms and surface plasmons. *Nature Physics.*
45. Szafron, R., & Czarnecki, A. (2016). High-energy electrons from the muon decay in orbit: Radiative corrections. *Physics Letters B, 753*, 61–64.
46. Tame, M. S., et al. (2013). Quantum plasmonics. *Nature Physics, 9*, 329–340.
47. Ting, T. C. (2004). The polarization vector and secular equation for surface waves in an anisotropic elastic half-space. *International Journal of Solids and Structures.*
48. Waseem, S. B. (2011). Orbital excitation blockade and algorithmic cooling in quantum gases. *Nature, 480*(7378), 500–503.

Index

A
Advanced building design, 145
Air dynamics transport and energy state travel machine, 164–167, 171–172
Air quality, 14
Air water molecule, 79
Alternative energy source, 65

B
Breakdown of CO_2, 177
Breathing problem, 18

C
Carbon (CO_2) emission, 3–6, 8, 13–18, 22, 24, 39, 189
Chemotrophic mechanism (Ct_m), 182–185, 190
Climate change, 3–6, 8, 13–18, 21–26, 29, 37, 39, 49, 65, 75, 87, 113, 122, 127, 135, 151
Climate change impact on ecosystem, 21
Climate change mitigation, 23
CO_2 breakdown, 39
Cognitive power (C_p), 91–94, 100–102, 107
Conventional industrial development, 14, 18
Cross approach analysis global ecosystem, 21
Curtain wall, 75, 113–115, 117–120, 122, 123, 127–129, 134, 135

D
Dormant light energy (D_e), 39
Drinking water, 154

E
Electric fields (E_e), 29, 32, 37, 63, 151–153, 155, 161, 167, 168
Electromagnetic fields, 33, 47, 57, 91, 96, 102, 107, 113, 115, 117, 119, 123, 152, 159–163, 167–169, 171, 187
Electromagnetic force of the earth, 161
Eliminating global pollutants, 177
Environmental energy (Ee), 91, 95–99, 103, 107
Extreme relativistic condition, 69, 74

F
Finite level of fossil fuel reserves, 135
Fossil fuels, 55, 65, 113, 127, 135

G
Global CO_2 equilibrium, 29–37
Global ecosystem modeling, 21
Global health crisis, 3–6
Global sequestration, 7

H
Helium-assisted curtain wall, 114, 122
Higgs–Boson BR ($H\rightarrow\gamma\gamma^-$) quantum field, 113, 119, 121, 123
Higgs–Boson ($\mathrm{H} \rightarrow \gamma\gamma^-$) magnetic field, 91, 96
Higgs–Boson's electromagnetism, 164, 165, 171, 172
Hossain ultrasonic germicidal frequency (HuGF), 139–146

© The Editor(s) (if applicable) and The Author(s), under exclusive license to Springer Nature Switzerland AG 2025
M. F. Hossain, *Sustainable Global Development in Environment, Energy, Building, and Infrastructure*, https://doi.org/10.1007/978-3-031-84429-4

Hydrogen energy, 86, 87
Hypercharge electrons, 42–44, 180–182, 193
Hypercharge striking electron, 182

I
Invisible infrastructure, 159–172

K
Killing pathogens, 141–142, 144–145

M
Mechanistic climate modeling, 22–24

N
Natural cooling and heating mechanism, 113
Non-equilibrium photon production, 70, 74, 75

P
Photo-electrolysis, 81–85
Photon energy, 32, 34, 60–62, 79, 117, 119, 122, 123
Photon frequency (f) and wavelength (λ), 31, 32, 34–36
Photonic band gap (PBG), 60, 61, 71, 73, 113, 114, 117
Photosynthesis, 29, 151
Phototrophic energy (Pt_e), 177–182, 185, 193
Physical energy (P_e), 91, 94–96, 102, 103, 107
Political barriers, 3–6, 8
Population density, 3
Printable solar cells, 127–135

Q
Quantum electrodynamics (QED), 49, 91, 94, 102, 103, 107, 129, 180, 193

R
Renewable energy, 55, 69–75, 127
Renewable energy technology, 69–75
Risk estimates from climate change, 21

S
Solar energy, 32, 35, 42, 43, 45, 55–65, 69, 71, 73, 113–123, 127–129, 132–135, 154, 178
Sonic frequency, 143
Spherical harmonics, 159
Static electricity, 80, 151, 152, 154
Static electricity force, 79, 84, 86, 87, 155

T
Thermal fluctuation, 61
Transpiration process, 152
Transportation sector (TS), 13, 14, 18

U
UV technology, 154

W
Water molecule, 79–87, 151, 152, 156
Water tank, 80, 81, 83, 86, 151, 154
Water tugging, 152, 154, 155

The manufacturer's authorised representative in the EU is Springer Nature Customer Service Centre GmbH, Europaplatz 3, 69115 Heidelberg, Germany. If you have any concerns regarding our products, please contact ProductSafety@springernature.com

Printed and bound by CPI Group (UK) Ltd, Croydon, CR0 4YY
01/10/2026
02225980-0001